Farm Life

A Century of Change
for Farm Families and Their Neighbors

Based on a Major Exhibition

Text by
Frank Smoot

Funded in part by the
National Endowment for the Humanities

Chippewa Valley Museum Press

Chippewa Valley Museum • 1204 Carson Park Druve • PO Box 1204 • Eau Claire, WI 54702

Printed and Bound in the United States of America

CVM *Project Development Team*
Sara Anderson, director of public programs
Karen DeMars, educator
Janet Dykema, director of community programs
Dondi Hayden, facilities manager
Susan McLeod, director
Jeanne Nyre, designer
Carrie Ronnander, curator
Kathleen Roy, assistant curator
Susan Sveda-Uncapher, assistant designer
Frank Smoot, director of publications
Eldbjorg Tobin, librarian

Artifact photography: Kathleen Roy

Cover and book design: Frank Smoot

This book is published as one component of
Farm Life: A Century of Change for Farm Families and their Neighbors,
a project that includes a gallery exhibit at the Chippewa Valley Museum
and other programming.

Funded in part by a major grant from the National Endowment for the Humanities.
Any views, findings, conclusions or recommendations expressed in this publication
do not necessarily reflect those of the National Endowment for the Humanities.

Library of Congress Catalog Card Number: 2004095053

Smoot, Frank. *Farm Life: A Century of Change for Farm Families and Their Neighbors*

ISBN# 0-9636191-4-4

Contents

Preface

Frances Jane (Hobbs) Pierce, seeding a pasture with a drill and four-horse team, 1942. Courtesy of US Department of Agriculture (USDA), Natural Resources Conservation Service. Frances Hobbs married Marshall A. Pierce in 1928. She was a partner in the farming operation until about 1955 when she returned to teaching in the Hillview school, Town of Washington, and at Augusta elementary school. The Pierces retired in 1968 and moved to Adams Co. where they lived until their deaths in the 1980s.

The photo is one of a series taken by the USDA, promoting soil conservation after the dust bowl years in the 1930s. The Pierce farm was one used as a model for contour farming and pasture regeneration. This photo was used to demonstrate how a grain drill can be used to reseed overgrazed or neglected pasture land.

The buildings in the background were those of William and Hilda Schlewitz. The field was located in the southeast quadrant of the intersection of Mayer Road and Prill Road. The Schlewitz buildings were still standing in 2004, but the pasture land had become a tract of homes.

Robert Lotz grew up on a farm in Chippewa County. A natural storyteller, he readily tells about his life as a child: driving his uncle's team back from town, his city-born mother's very bad luck with mail-order chicks, and his dad adding a rural mail route to his work on the farm. As an adult, Robert Lotz became a surgeon in Eau Claire, but history was his special interest and he helped found the Chippewa Valley Museum (CVM). In the late 1970s, he spearheaded a drive to add a new wing dedicated to the region's rural heritage. With broad community support and hours of volunteer effort, the addition opened in 1982. Twenty-two years later, with an expanded research and collections base, CVM installed *Farm Life: A Century of Change for Farm Families and Their Neighbors*, a major new exhibit for the Rural Heritage gallery. This publication is a companion to the exhibit, which tells a story of profound change for farm families and rural communities.

The study of agricultural history and rural experience provides important insights into the evolving character of American life. Our region presents an intriguing vantage point for this consideration. As a whole, the area is distinctly rural. In 2000, about 344,000 people lived in the 9,500-square mile Chippewa Valley. With just over 60,000 residents, Eau Claire (home of CVM) had four times the population of the next largest city. Agriculture and agricultural processing industries have been economically important here for more than a century and remain so in 2004 despite rapid change. Transitions occurring in the Chippewa Valley now parallel national experience elsewhere twenty or thirty years ago.

The new exhibit needed both a strong research base and a distinctly regional voice. We found our research methods in the work of folklorists and regional writers and in new historical scholarship that focused on the importance of family and community in rural life. To find the voice, in 1998, CVM initiated an oral history project, *Fields and Dreams*. CVM staff, consulting historians, and trained volunteers (five of whom were farmers themselves) interviewed 85 residents from farming or other rural backgrounds. Through the Rural Life Documentation Initiative, folklorists created deeper studies about regional people and places.

Between 1999-2002, CVM received three grants from the National Endowment for the Humanities to plan and build the new exhibit, create related publications and programs, and develop resources for educators and other museums. The Wisconsin Humanities Council, the Wisconsin Arts Board and the National Endowment for Arts also supported various aspects of these activities. CVM is deeply grateful to these agencies and to the farmers, writers, scholars, artists, museum and multimedia specialists, and others who have made our work and these results possible.

— Susan McLeod, Director, Chippewa Valley Museum
September 2004

Foreword

In my house there is a book that breaks my heart. Bound in burnt orange hardback, it hefts like a slab of bacon. Opened, it covers your lap. It came in a box with other books, which a friend bought off a decrepit hay wagon at an auction. On the cover a weathervane is imprinted in black silhouette next to the title, *Down on the Farm*. Published in 1954, it is filled with black-and-white pictures of farms then fresh and sturdy: The white clapboard house with its first coat of paint, the red barn rawboned and square, with nary a sag nor patch of tin. When these pictures were taken, the farms were not history, they were the future. Looking at those pictures today, you figure there's a good chance the troweled foundations were long ago reduced to tumbled traces in some sumac patch. Early in the book there is a picture of a farmer walking up a grassy two-track toward his trim homestead. The path is flanked by elms, and the farmhouse is framed in sun-shot leaves. I long to kick along in the dust beside the farmer. It's just as well this is not possible, because I would feel compelled to warn the farmer that the world was gaining speed. For all I know, he would tell me it wasn't gaining speed near fast enough.

I knew years ago that I didn't want to milk cows for a living, but when I time-travel in my head, our old milk barn is a favorite destination. On those Wisconsin winter nights so clear and cold you could tip your head back beneath the yard light, puff out your breath, and fog up the stars, stepping into the cow barn was like dropping into a quilt. I can conjure the warmth of the Holstein's flank, the scrape of the three-legged milk stool when I hiked it beneath my haunches, and the feel of the straw beneath my boots. The milker went chit-chhh—chit-chhh—chit-chhh and there was no better cocoon in the world.

I'll stop there, because my job here is to introduce a story told by many voices. I'm happy to report that the people you will meet in *Farm Life* don't just mope around staring all moon-eyed at pictures in an old book. They tell a story that is part celebration, part history, and part elegy. But for every wistful memory, you get a clear-eyed report on the gritty aspects of farm life we tend to gloss over in our gilded recollections. And you get humor. I am thinking in particular of the anecdote about the threshing crews pranking on the bundle hauler by tying the last grain shock to the wagon. I grew up on that manner of rough-hewn practical joke. It was homespun humor, and it made the endless chores go better.

Farm Life tells the story of Chippewa Valley farming clearly, honestly, and best of all, with heart. I make that claim from just one humble point of authority: I was raised a farm boy of the Chippewa Valley.

— Michael Perry, New Auburn
October 2004

Notes to the reader:

1. Many kinds of animals, and many kinds of crops, foodstuffs, and other plants fill out the rich tapestry of farming in Wisconsin's Chippewa Valley.

   Farmers raise, or have raised, dairy and beef cattle; dairy, meat and wool goats; horses, sheep, swine, llamas, buffalo, red deer, and mink; and, turkeys, chickens, ducks, and geese; among other

Above: Oak Grove Berry Farm, Town of Union, Eau Claire County. Courtesy of George Johnson.

Right: Hmong gardeners sold summer produce at the Downtown Farmers' Market, Eau Claire, August 1994. Photographer: Jason Tetzloff.

animals. They've grown wheat, barley, oats, and other small cereal grains; timothy and alfalfa hay; field, sweet, and Indian corn; potatoes, yams, and other tubers; soy, lima, navy, red kidney, snap, and other beans; asparagus, beets, cabbage, carrots, celery, cucumbers, onions, peas, pumpkins, squash, tomatoes, and other commercial vegetables; apples, cherries, strawberries, melons, and other fruits; and, horseradish, rutababa, ginseng, and other roots. People have produced honey and maple syrup. The area's early farm settlers grew tobacco as a cash crop. In recent decades, the Hmong, immigrants to the Chippewa Valley from Laos beginning in the mid-1970s, have adapted their centuries-old agricultural tradition to engage in "truck gardening," selling fresh produce at area farmers' markets.

But for a century, dairy has been the area's main agricultural claim to fame. Wisconsin is known as America's Dairyland, and the Chippewa Valley community of Eau Claire was mandated for decades by federal regulation (the "Eau Claire Rule") as the geographic center of U.S. dairying. And so we look at our story of farm life through the dairying detail of this tapestry.

2. Most of the people quoted in this book are, or were, members of farm families. Volunteers and staff at the Chippewa Valley Museum (CVM) interviewed more than eighty individuals between 1998 and 2004, whose stories became the *Farm Life* exhibit and book. Unless otherwise noted in the attributions, you are reading the words of Chippewa Valley, Wisconsin, farm family members.

3. Unless otherwise indicated, all photographs, documents, and artifacts reproduced in these pages are found in the CVM collections.

Why?

Farmer wins the lottery. Reporter asks him what he's going to do with the money. Farmer says, "Keep farming, I guess, 'til it's all gone."

In 2000, Harold Kringle lived near Barron, and he recalled how he came to be there in the first place. "When I came back from the service, it was back in September 1945," he said. "And on the way home, we went to Madison to look into going to the university there, and then we came back up here. At that time this farm was for sale.

"Carol's dad said, 'What kind of work — do you want to farm?'"

"I guess I said, 'No, I want to go to school first'."

"Then he said, 'Well, there's a "school" for sale, so let's go look at it'."

He and Carol bought and started farming it. They were dairy farmers — milking as many as sixty cows — although they raised some beef cattle and soybeans. They raised corn and alfalfa as feed for their herd. Their son took over the operation and Harold retired "about an hour ago."

"If I would have gone to the university —" He paused to reflect. "But I think I still would have wanted to farm after I got out of school. I don't know if it is insanity or what."

Anyone who gets a few chapters into any history of farm life might wonder the same. Why do farm families farm? They mention a love of the land, of working outdoors, of variety, of independence. "If there's an auction someplace at one o'clock," said Gary Evans, interviewed in 2000 at the Chippewa Valley Technical College in Eau Claire where he taught farmers, "they can go to it. You know, they don't have to ask a boss if they can go." The answers that Chippewa Valley farmers gave have a common thread, and that thread is about how each day goes, and how one day flows into the next.

LaVerne Ausman recalled calm evenings in his childhood in the 1940s. "Behind the house ... and around the barn and granary ... in the summer, we would have a single wire kind of strung around all of this, and we would turn the horses in there before we let them out to pasture, and then we would sit on

Jane Mueller helped her son Steven learn how to feed a calf at the Mueller farm near Fall Creek, 2000. Photographer: Jeanne Nyre.

the back steps of the house and talk and maybe have homemade root beer, maybe some ice cream, and watch the horses eat. Then when they would ... start getting restless ... you would chase them out and send them out to pasture, and that was the end of the evening."

For Bernice Sutliff, the question *why farm* brought to mind the air moving above her. "Just standing up on the hill with the wind blowing or, you know, just looking out over your cattle or over your land and, and the windmill whipping around above you, I mean it's, it's a good feeling. Beautiful, beautiful ... And I don't know ... I'm sure there were some bad things that happened. But ... it's just, just a naturalness that these things happen, and you have to be ready. But it's, it's, it's a beautiful life."

In 2000, Jan Morrow, who with her husband bought her farm between Cornell and Bloomer in 1993, cast her answer with a thought towards the end of farming, as if, even on their recently purchased property, the "family farm" was in its autumn. "We get up together every morning and we go to bed together every night, and we are together ... all day long. And with the plight in farm economy right now, you know, there is the point, are we going to have to sell the cows, and is my husband going to have to get a job.... It's not that I can't manage without him; it's the idea that I'll see something on TV or I'll hear something on the news, and I'm so used to being able to walk down to the barn and tell him."

On the other hand, the Mueller family, interviewed the same summer, cast everything very much in the present. Jane and Doug Muellers' three children, Meg, Peter, and Steven, all helped out with the canning. Steven, seven at the time, helped with mashing the peaches, and everyone took turns cranking the sieve to turn the tomatoes into a soup base. Meg, seventeen then, had gotten an off-farm job and did miss milking with her father every night: "Probably one of my favorite parts when I was milking was that I got to spend almost two hours with dad every night, and we could talk about whatever. I miss that a lot, because now, especially with work, he's already in the barn when I go to school and he's in bed when I get home from work.... But I think I do have a stronger family relationship than a lot of people I know, a lot of people I go to school with."

Many people can see the value of *the country*, away from the crowds and roads and pace of city life, and many people have bought forty acres *not* to farm it. In fact, since forty acres qualifies a plot as a farm, the number of small "farms" in St. Croix County (the Wisconsin county closest to the Minnesota Twin Cities of Minneapolis and St. Paul) actually increased between 1997 and 2002. But farm families will tell you that *farming*, trying to make a sizable chunk of one's living off the land, is in the blood and not the mortgage. It's what makes the physical labor and frequent economic uncertainty not evils to be overcome or avoided, but complexities that make the life more deeply and richly sustaining. "And if you understand this," Emil Sarauer said on his farm near Bloomer, "this is heaven."

Foundations

The great dark trees of the Big Woods stood all around the house, and beyond them were other trees and beyond them were more trees. As far as a man could go north in a day, or a week, or a whole month, there was nothing but woods. There were no houses. There were no roads. There were no people. There were only trees and the wild animals who had their homes among them.

— Laura Ingalls Wilder, Little House in the Big Woods

Lars and Grethe Anderson, 1880s

The Happy Yeoman

In 1847, German immigrant George Meyers settled on land near Tilden, Wisconsin, and had seed and field-working gear shipped in by boat up the Chippewa River. Crops had been grown in the Chippewa Valley for 1,000 years, but tradition holds that he was the first white farmer in the Chippewa Valley, and as such, he was also bringing the new settlers back into the folds of an ancient human tradition.

Theodor de Bry. Florida Indians planting beans or maize. *Engraving after a watercolor by Jacques Le Moyne de Morgues, 1564. From de Bry,* Brevis narratio eorum quae in Florida Americae. *Library of Congress, Rare Book and Special Collections Division. Digital ID: LCUSZ62-31869.*

Farming as an Occupation

Myers was pursuing an occupation that dates back more than 10,000 years. At that time, in what is modern-day Iraq, people maintained herds of sheep, while in Kurdistan, they may have been planting wheat 1,000 years earlier yet; that is, saving part of their harvest to scatter where they gathered it.

Meanwhile in Mexico 9,500 years ago, neolithic farmers were selecting those seeds from a non-descript wild grass, *Balsas teosinte*, that produced the most tightly knotted clump of nutritious seeds on a cob. Through this selection, over successive generations, these farmers genetically molded the grass into maize.

Taming animals, grains, and fruits allowed people to stop wandering the earth in search of food. They settled onto parcels of land and into villages and cities, which became marketing hubs for foodstuffs that the farmers raised, and manufacturing centers for goods made of farm products, such as wool, rope, and papyrus, an early form of paper. In this way, agriculture created urban culture.

In the lower portions of the Chippewa Valley, groups were engaged in agriculture from about 1000 A.D.., primarily growing corn, beans, and squash, known as the "three sisters." They were complementary crops, grown on the same mound: the corn stalks served as trellises for the bean plants, the large leaves of the squash shaded the ground and helped it retain moisture, and minerals in each of the three plants replenished minerals the other two took from the soil. The three sisters did not replace hunting, fishing, and gathering, which people had already been doing in this region for 5,000 years; they supplemented those wild foods. Woodland villages, often numbering 200 people or more, were mobile. Villagers farmed open land in the summer, hunted the woods in the winter, and fished the waters in the spring. Men burned the brush off 100-acre fields, and women grew the crops, which supplied most of the calories the Woodland villagers consumed. But, some time around 1200 A.D., for reasons not yet known, village farming disappeared in the Chippewa Valley, although hunting, fishing, and gathering continued. In other nearby areas, including southern Wisconsin, village farming continued.

Woodland chert hoe, 1000-1200 A.D.

Ojibwe Culture

The Ojibwe, who came to Northern Wisconsin in the 1700s from the eastern United States, were not village farmers in the same sense the Chippewa Valley's ancient Woodland people were. Still, in addition to hunting and fishing, they not only gathered wild rice but sowed and managed it.

The Ojibwe at Lac Courte Oreilles told ethnologist Albert E. Jenks that "rice was sewn in an easterly direction from one body of water to the next until it reached every Ojibway community." Rice was gathered from Red Cedar Lake to seed Lake Chetek, Rice Lake, Bear Lake, Moose-Ear Lake, and the Lac Courte Oreilles River. The latter was seeded about 1860, after the 1854 treaty set reservation boundaries.

Harvesting wild rice. Drawing by Seth Eastman, c. 1831-32. Courtesy of the Wisconsin Historical Society. WHi (W6) 5938.

Although wild rice gathering is an iconic Ojibwe activity, it was mostly taken away from them by white settlers, who argued that they could make wild rice beds more productive. Producers mechanized sowing, harvesting, winnowing, and scorching wild rice. Processing plants were built, and by 1900, dealers began advertising the product.

European Village Farmers

Around 1000 A.D., the same time the Woodland people were pursuing agriculture in the Chippewa Valley, the first European villages formed under the leadership of large landowners. To peasants who lived on and worked the manoral land, their lords offered protection against the anarchy and starvation outside their manor walls; in exchange, the peasants were obligated to them for life. While better than the alternative, life on the manor was not happy for the peasants: it offered them unending days of dull, back-breaking labor with no hope of bettering themselves. This manoral system lasted until the bubonic plague swept Europe around 1350. In a harsh irony, the Black Death improved the lot of the peasants who survived it. It caused a labor shortage that allowed peasants to set their own terms for working the land. Some abandoned the manor altogether; others stayed on with pay, something unknown even to their most distant ancestors.

Farming near Vienna, Austria. Courtesy of the Wisconsin Historical Society. WHi (X3) 30443.

They formed villages of their own. Five hundred people or more might live in a village. These were still rigid societies, dominated by those with the most money and land. But everyone had work, and it was work in the service of the village. In one pattern of settlement, cattle grazed "common fields," anyone could cut his or her fuel from the woods, and all those who were able helped build and maintain village roads. Even though individual farmers owned land, the village decided which crops they should plant. If a person lacked work, the village assigned work to him. Like their Woodland counterparts across the Atlantic, European villagers believed in the welfare of their fellow villagers.

This particular mode of European village farming lasted relatively briefly. By 1500, the strengthening of nation-states diminished village autonomy. Travel and trade offered new ideas, methods, and markets; exotic goods to desire; and opportunities for independence and land ownership in whole new worlds. But the village idea of mutual assistance survived.

By the eighteenth century, Europeans had made agriculture into a complex and well-developed science, as it had also become in other parts of the world. European farmers used plows to break the ground, rotated crops to keep the soil from wearing out, and were trading agricultural commodities both with neighboring villages and far afield.

Farming as an Idea

In starting his farm near Tilden, George Meyers was pursuing an idea as well as an occupation. In the western world, which lent its ideas to modern Chippewa Valley farming, the idea that farming is a noble occupation dates back before the Rome of Cicero and Cato, at least to the Hebrew prophets. Shemaiah advocated a work ethic, believing there was great dignity and inspiration in manual labor, such as farming. In the Bible, farming begins as a curse. In the Book of Genesis, as God banishes Adam and Eve from Paradise, he tells them that their life of leisure is over, that by "the sweat of your face you shall eat bread." But by the Book of Isaiah, the curse had become a blessing. Even though God "gives the bread of adversity and the water of affliction," those who walked in the way of the Lord would have rain for their seed crops, a rich and plenteous harvest, and their cattle would graze in large pastures.

By the time the western world reached 1776 and the English were well established in America, the work ethic had grown into a well-articulated political stance "that all men are created equal" and that the most basic rights were life, liberty, and the pursuit of happiness. But hard work was not enough; hard work was serfdom or enslavement without the freedom and independence brought by owning land. The egalitarian occupation of yeoman farming perfectly fit the ideals of the new American republic. The sweat of one's brow was honest payment and brought a measure of self-sufficiency and independence never even dreamed of by Medieval peasants.

> *Every blade of grass has its angel that bends over it and whispers, "Grow, grow."*
> — *The Talmud*

The Agrarian Ideal

Yeoman is an ancient word, used by Chaucer, derived perhaps from the same root as the word Goth, meaning "the good people." Yeoman has several meanings, but in the sense of farmer, a yeoman is a free person who cultivates his own land.

America's founders paid no higher praise to anyone than to the yeoman farmer. In *Notes on the State of Virginia*, Thomas Jefferson wrote, "Those who labor in the earth are the chosen people of God, if ever he had a chosen people, in whose breasts he has made his peculiar deposit for substantial and genuine virtue."

Above: Yeoman Farmer. From Virgin Land: The American West As Symbol and Myth *by Henry Nash Smith (1906-1986), a Synoptic Hypertext, American Studies Group, The University of Virginia 1995-96.*

Below: Sharpening scythe, Richland County, Wisconsin, July 1941. Photographer: John Vachon. Library of Congress, Prints & Photographs Division, FSA-OWI Collection. Digital ID: fsa 8c36155. Source: intermediary roll film.

This is the agrarian ideal, a word that comes from the Roman *agrarius*: the collection of laws, or the advocates of those laws, favoring the division of public lands among Rome's poorer citizens. Agrarianism has influenced western culture since classical times, and has informed America's idea of itself to the present day. Agrarian thinking is central to the American Romantic movement in literature (Walt Whitman is one example of an American Romantic), Transcendentalism in philosophy (Ralph Waldo Emerson and Henry David Thoreau) and Populism — a political philosophy, movement, and party, supporting the rights and power of the common people against the elite. While short-lived, Populism nonetheless set in motion many social changes in the twentieth century.

Jefferson believed what he said, but it also served two important political purposes. The idea that independent white adult males could be full citizens was central to the American system of government. (Over many decades his definition, generous at the time, expanded further to include women and people of other races.) And, the idea that a simple, harmonious, rural life was more virtuous than an intellectual, discordant, urban life was essential to a healthy American self image: In 1790, 90 percent of American households were farm households.

> *Through the ample open door of the peaceful country barn,*
> *A sunlit pasture filled with cattle and horses feeding,*
> *And haze and vista, and the far horizon fading away.*
> —Walt Whitman, "A Farm Picture," from Leaves of Grass, 1891

Gone But Not Forgotten

> *What, then is the American, this new man? ... Wives and children, who before in vain demanded of him a morsel of bread, now, fat and frolicsome, gladly help their father to clear those fields, whence exuberant crops are to arise to feed and to clothe them all; without any part being claimed, either by a despotic prince, a rich abbot, or a mighty lord ... From involuntary idleness, servile dependence, penury, and useless labor, he has passed to toils of a very different nature, rewarded by ample subsistence. This is an American.*
>
> — Hector St. John de Crèvecoeur,
> Letters From an American Farmer, 1782

Ironically, by the time the Populist movement reached its height in the Farmer's Revolt of the 1890s, American social and technological change had made the farmer not a yeoman anymore. He bought land

and machinery on credit. He transported his goods nationwide. He joined national organizations such as the Grange and the Farmer's Alliance. He was not the self-sufficient husband of the land doing God's work in solitude. He was intimately involved in the U.S. economy, and he was subject to its ups and downs like any other businessman. By the time George Meyers settled near Tilden in 1847, America was well on its way to making the transition from the life of farming to the business of agriculture.

In 2004, farmers made up 2 percent of the U.S. population, and those farm families that remained were caught between conflicting perceptions: to be virtuous but not unsophisticated, independent but not alienated, astute business agents but not greedy recipients of subsidies or rapacious abusers of the ground and groundwater.

> The new sustainable American farmers are "family farmers," in the truest sense. They are stewards of the land, they value relationships, and they are pursuing a more desirable quality of life — economically, socially, and spiritually…. Can America depend on these new farmers? We can if we make it possible for them to remain true family farmers, sustainable farmers, instead of forcing them to exploit the land, their customers, and each other in vain attempts of economic survival. These new farmers are real people…. They are rooted in the place where they grew up, where they have family, and would like their children to "take root" in those places as well. They are Americans.
>
> — John Ikerd, Professor Emeritus
> of Agricultural Economics, University of Missouri,
> "Farm Economy State of the Union Address," 2002

Complicating the Story

American farmers were never completely independent, of course. They received important assistance from the federal government. The federal government assumed responsibility for negotiating treaties with American Indians that gave white settlers the legal title to land in the Chippewa Valley. The government then distributed this land, often to speculators and dealers, but sometimes directly to farmers through low-cost programs such as the Homestead Act. It also subsidized the roads and railroads on which farmers moved their supplies and goods.

On the other hand, it has been a relatively rare time since the 1890s that American farm families were "fat and frolicsome," as de Crèvecoeur declared them in 1782. Economic anxiety, physical and environmental hazards, and isolation have been companions of farm families. Foreclosures, suicide, and domestic violence have been occasional unwanted guests. Decisions farmers have made, especially about credit, have played some part in their financial misfortunes.

While farm families are subject to the same economic stressors as any other group in the U.S. economy, they are also subject to hardships out of their control. The farm crisis of the mid-1980s was brought to the attention of non-farm families by singer Willie Nelson. It was followed closely in Wisconsin by the drought of 1988, feed shortages in 1989, low milk prices in late 1990 and 1991, drought and alfalfa winter kill in 1992, floods and alfalfa winter kill in 1993, and extreme heat in the summer of 1995. Shortly after, milk prices started dropping again, falling to $10 per hundred pounds of milk in 2000, prices dairy farmers hadn't seen since the 1970s. From 1987 to 1997, Wisconsin lost 23 percent of its medium-acreage farms, those between 180 and 500 acres.

> *The farms are ploughed under, bulldozed, erased.*
> *The trees are gone. The creeks diverted. Grass*
> > *where the slopes were thrown is still fire-green.*
> *My parents are dead.*
> > — *Robert Peters, from his poem "Now" in the collection* What Dillenger Meant to Me.
> > *Peters lived in Vilas County in the 1930s.*

A Reason to Believe

The twentieth century was a century of change for farm families and their neighbors. In 1900, 42 percent of the American population worked in agriculture. At the end of the century, about 1.8 percent were involved in farm production — so few that in 2000, the U.S. Census Bureau considered removing them as a separate occupational group. In 2001 in Wisconsin, almost 100,000 people worked in farm production. This was 3 percent of the state's population, which beat the national average, but still the number had dropped by half since 1981.

Even in Wisconsin's deeply rural Chippewa Valley, most people are at least two generations removed from farm life. Nationally the economic viability of the family farm has been weakened by the growth of corporate agriculture — although, in the Chippewa Valley at least, most large farms aren't faceless multinationals. They've often developed from combinations of smaller operations, parents and children, or neighbors. Those families, and others on single-family farms, have retained their connection to farming

— and even particular farms and rural neighborhoods — sometimes through many generations. In 2004, farm life still compelled one family to stay while a score of others depart.

Is it important to the rest of us? There are reasons to think so. First, as the saying goes, "If you eat, you're involved in farming." In the worst case, if farming were finally taken offshore for the sake of profitability, our food would be subject to the same jeopardy as our crude oil. Less fundamentally, but still compellingly, the journey of the yeoman farmer rides in tandem with the journey of the American citizen, the farm family with the American family.

Sandy Acres Dairy, Jeff and Marie Pagenkopf's farm near Elk Mound, 2002. Inset: The same farm, owned by Walter Pagenkopf, in 1956.

Into the Valley

Some came from Europe. Others came from American cities. Some came to ply the lumbering trade; others only used that trade as a means to earn cash for land. Among them, they cut enough trees off the land to fill the American Middle West with houses, taverns, churches, and schools. Still others came with an eye only to farm a small parcel of the "land of cows and clover," the six million acres of the Chippewa River watershed.

Dells Lumber Company employees near Kennedy, about 1910.

The Pinery

The men would go into the woods in the fall, then come back in the spring with their money… and that was the only means of getting any money, really… there wasn't much dairy in them days. You never milked in the wintertime… when I was a kid, anyway…
— Carl Penskover, Rice Lake, interviewed 1998, speaking of the 1910s and 1920s

The signing of the 1837 "Lumberman's Treaty" between the United States and the Ojibwe opened the great white pine forest of the Chippewa Valley to lumbering. Yankees, Canadians, Germans, and Irish poured in, followed quickly by Scandinavians.

In October 1854, 13-year-old John Barland and his 10-year-old brother Thomas each drove an ox-wagon of potatoes to a logging camp on the Wolf River, a tributary of the Eau Claire. They forded the Eau Claire on the backs of the oxen. Three months later, in the dead of winter, John almost froze to death after falling in the river on a return trip from the camps.

Local farms such as the Barlands' supplied some needs of the men and animals in the logging camps. Proprietors of "stopping places" along the logging trails — where loggers could get food, lodging and feed for their animals — also did some farming, and sold their harvest to the camps. But neither group could keep up with the demand. Thousands of men, and countless draft teams, worked the Chippewa pinery at its peak in the 1880s.

Because local family farms couldn't supply the population surge — and importing grain, butter, cheese, and meat was expensive — the logging companies started their own farms. Daniel Shaw's 900-acre Flambeau Farm supplied his lumber company with vegetables, grains, meat, hay, and draft animals. But even that farm, huge by the standards of the day, was modest compared to the six farms, comprising 6,000 acres, operated by Menomonie's Knapp, Stout & Company. Prairie Farm, still an active community in 2004, was one of those places.

The Plow Follows the Axe

Within two or three generations of the first influx of loggers, the Chippewa Valley's "inexhaustible" pinery, holding 46 billion board feet of white pine lumber, had largely disappeared.

By working in the winter camps, many immigrant men had earned cash for land, tools, stock, and seed, while women and children maintained the farms in their absence. Once the land was useless to them, lumber barons and land speculators sold the "cutover" at bargain prices to farmers, many of whom were their former workers.

Cousins playing on a tire rim swing on a farm that had been woods only a few years earlier, eastern Taylor County, 1938. Left to right: Robert Cooper, Inez Cooper, Irene Cooper, Herbert Cooper. Courtesy of Inez and Roger Robertson.

Other newcomers had been aided by the Preemption Act of 1841, which allowed farmers to settle on unowned land. Those who agreed to farm the property, improve it by building, and declare an intention to become citizens, could buy up to 160 acres for $1.25 an acre, if they paid for it within a year of settling.

The Journey to America

On April 20, 1853 — three weeks after they were married — 27-year-old Lars Anderson and 22-year-old Grethe (Paulson) Anderson left Christiania, Norway (now Oslo). They sailed to America accompanied by Lars' younger brother Jens.

At the time, Norway was a land of very little opportunity. More than a million people lived in Norway and three-quarters of them were trying to make a living from the land, less than 5 percent of which was tillable. There was virtually no land left to homestead, and much of the land that might be inherited was mortgaged beyond any young family's ability to pay.

The Andersons spent seven weeks at sea. Under the best circumstances the trip was miserable. Overcrowding was severe, and seasickness was universal. On voyages like theirs, typhoid, dysentery, and measels claimed scores of lives. But on June 7, Lars, Grethe and Jens arrived safely in New York, three of more than 6,000 Norwegians who reached the ports of New York and Quebec in 1853 alone.

On to Wisconsin

The Andersons traveled to Milwaukee by lake steamer and walked to Waupun, 50 miles to the northeast, a common destination for Norwegian immigrants. They stayed three years, working, saving money, and getting used to new customs and a new language. In the fall of 1856, carrying all their belongings and Lars and Grethe's one-year-old son Carl, they traveled by foot and covered wagon from Waupun to Eau Claire, a journey of more than 150 miles.

Trunk belonging to Lars and Grethe Anderson, 1853.

They wintered on the banks of the Chippewa River and lived the next year in what is now the town of Pleasant Valley. The following spring, Lars and Grethe moved with Jens, and a number of friends from the Old Country, to Chippewa Falls township, where they would all live out their lives.

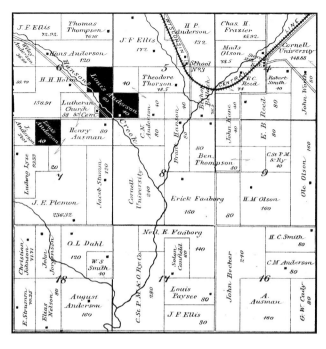

[Waupun was] a stopping for all [Norwegian] immigrants bound for the West, as a great number of old settlers were residents of that city at one time.
— Mrs. Ole Tilleson, who came to Elk Mound from Waupun in 1862. Quoted in the Eau Claire Leader, November 12, 1911.

Detail of plat map, Town of Wheaton, Chippewa County, 1888.

Staking a Claim

Lars and Grethe Anderson selected 120 acres in Chippewa Falls township. (It split in 1860, leaving their land in the Town of Wheaton.) They chose acreage with a handy water supply. Big Elk Creek cut across a corner of the farm. The rolling terrain was mostly prairie brush with a stand of pine and a dotting of hardwoods.

Lars had made a "preemptive" claim on the land as soon as he arrived in the area. His brother Jens shortly claimed an adjacent tract. The Anderson's weren't able to pay, and the land went to a Chippewa County land speculator named Andrew Moore. Lars and Grethe never moved, however, and Moore sold the farm to them in 1861. This chain of events, which allowed farmers extra time to get money together and made profits for speculators, was not technically legal, but it was a common practice.

The Andersons cleared their land by burning the brush cover and pulling stumps, then broke the prairie sod with a plow and smoothed the field by pulling logs over it. Grethe brought her children out to the fields, where they could sit in the shade while she worked with Lars and Jens. Their first year on the land, the Andersons planted root crops and vegetables, which sustained both people and animals. Lars and Jens also hunted for wild game. The Andersons arrived in Chippewa County in the 1850s; 60 years later, the Slovenian-born Perovseks came to Clark County, only to face the same task.

The first year they just kept clearing the trees and cutting the wood ... and they'd seed a little grass between the stumps. That was 1908 when they came out — by 1910 they had a cow.
— Josephine Trunkel, Willard, daughter of Frank and Mary Perovsek, interviewed 2001

Little Farms in the Big Woods

It is a fact well established among buyers that this valley produces a better article of wheat than any other section of the state. Straight lots from this place brought 5¢ more per bushel than any other section last fall in Milwaukee.

— Eau Claire Free Press, *about 1863*

Neighbors taking a dinner break from cutting down pine stumpage to be used in the construction of Elmer Swan's new barn, Barron County, 1926. The building in the background was the Swan family's old milk house, moved to the wood cutting site to be used for shelter. Courtesy of Art Swan.

The land around Tilden was so good that George Meyers grew 27 consecutive crops of wheat from 1847 to 1874 without the use of fertilizer. But even though the soil was rich and the farmers prospered, farming was extremely labor-intensive and removal of the giant white pine stumps proved next to impossible. Farmers planted by hand among the stumps, harvested with a cradle, bound by hand, and threshed with a flail, usually on the floor of the barn.

Frontier families typically milked one cow — "Jersey-types" were common — for family subsistence. A cow would often be the first major purchase a farm family made after their land. If enough ground could be cleared in a timely fashion to plant corn as feed, the farmer might buy a pig next after the cow. The pig would be fattened on corn and butchered in the yard on a cold day in the winter, so the meat would keep. The blood was kept for blood sausage, the hide for shoes, and the fat for lard (and to spread on bread when the cow was "dry," wasn't producing milk for butter).

Families might next get a draft horse or a team of oxen. A small stable or a lean-to attached to the side of the log house was often the only shelter for all these animals during the harsh Midwestern winters.

As a farm became more established, the farmer, his extended family, and the neighbors would build a separate barn structure. Still, the lean-to would find use as storage for the harvest and the few implements a farmer might have. A farmer who had gained some wealth would buy a wagon. It might well be that only then that the children would get to visit a city the size of Eau Claire.

Like most family farms of the era, Lars and Grethe Anderson's farm quickly became a diverse operation. By 1870, they had a team of horses, six milk cows, nine other cattle, five sheep, and five swine. In 1869, they raised wheat, Indian corn, oats, barley, potatoes, and hay, and produced 150 pounds of butter and 10 pounds of wool, according to census records.

There was a very small house and a log barn. Shortly after my great-grandparents moved there, they built the typical red Wisconsin barn. [But] the milking was done in the little log barn … [which] kept getting little additions … the calf barn, that was in the front … and the horse barn was attached to the back.

— Judy Gilles, Cadott, interviewed 2000. Her great-great grandparents came to the Chippewa Valley in the 1860s. In 1902, her great-grandparents bought the farm where she still lived in 2004.

Wedding coat worn by Lars Ronneberg when he married Antonina Anderson in the Big Elk Creek Church, May 11, 1895. Antonina Anderson was a daughter of Lars and Grethe Anderson.

Daily American Life

Lars and Grethe had ten children, but four of them died in childhood. And, between babies having yet to be born and young adults leaving home, a varying number of kids slept under their cedar shake roof at any given time. But in 1874, Lars, Grethe, and eight of their children lived in the three rooms and loft of their new 16' x 24' house.

The family spent most of its time outside or in the main room, where they cooked, ate, studied, bathed in winter, relaxed, conducted the business of the home and farm, and tailored and sewed their clothing. Many Norwegian settlers found that American-manufactured clothing became tattered under the strain of farm labor, but they quickly learned to tailor the clothes they made to an American cut. Tax records also show how quickly their names were Anglicized. Lars became Louis or Lewis. His brother Jens became James. On various documents Grethe became Greta, Gertrude, and even Margaret.

Go to a Yankee's house and you will be sure to see a tattered object workingaround, chopping wood, milking, or doing some such chore.

— Olaf Duess, 1856

Good Neighbors

No Norwegian settlement was complete until a church spire arose in its midst, and so was the desire of these pioneers.

— Big Elk Creek Church History, *1958*

The Andersons were soon surrounded by friends and fellow Norwegians, all originally from Baerum. The families of Gabriel Jensen, Hans Holm, Bernt Hanson, and Nels Hanson claimed nearby farms.

Lars' brother Jens donated the land for the school, on a property adjacent to Lars and Grethe's. This same land was later used as the site for the Big Elk Creek Lutheran Church and its church yard. Lars and Grethe even hosted the neighborhood's first church services until a formal structure coud be built. Building began on the church in 1876 and finished in 1884. It was still an active church in 2004. Lars served as church treasurer from about 1867 until 1894; his son Harold took over the job and remained as treasurer until 1941; they held the post 74 years between them. At various times, Lars also held positions as constable, justice of the peace, and town supervisor.

Big Elk Creek Lutheran Church, 1997.

Lars and Grethe's daughter Antonia was married at Big Elk Church in 1895, and many members of the family are buried in the cemetery. Rites of passage also took place in people's homes, especially before the church was built. After their second child Antomine died in 1864, her coffin lay in the house. After the service a procession led to the graveyard, where the men of the neighborhood took turns digging her grave.

Neighbors accomplished many tasks together. Although Gabriel Jensen was the craftsman who designed the Anderson's house and fit the joints, the Anderson's other neighbors lent their sweat. Home-building was a communal affair. House-raising bees, quilting bees and husking bees also provided relief from the isolation of pioneer life. Young men especially liked husking Indian corn. Anyone finding a red ear could kiss the girl of his choice.

Certain areas in the Chippewa Valley are still "mostly German" or "mostly Norwegian." The Connells and McIlquahams growing apples in Chippewa County in the 1990s were part of the "Little Ireland" settlement, which began in the 1850s and drew both Catholics and Protestants.

These farms, the names on the mailboxes, these are the same names.
—Irene Ovren, Albertville, interviewed 2001,
comparing the pioneer days to the present. Irene and Roald Ovren
lived in the Lars and Grethe Anderson's home briefly in the 1940s.

It Just Rains Rocks

It took us 15 years to clean up 25 acres of land for planting. The whole family worked at it.
  — *Reminiscences of John Mraz, Town of Lugerville, Price County, 1920s.*
  Phillips Czechoslovakian Community, Volume 1

Starting in the 1890s, a wave of immigrants — mostly Eastern Europeans — arrived to find the best farmland already taken. They clustered on the "cutover" land in the north, land that had been logged off and then all but abandoned by the lumber companies. Few of these families arrived straight off the boat. Most came from mines in Pennsylvania or Minnesota or the factories of Joliet or Chicago.

In the 1910s, the Barr family came to Willard after four years of drought on their farm in Montana. Lawrence Hart came from Illinois. Frank Perovsek was born near Grosuplje, Slovenia, but came to Willard from Chicago where he had been working in a steel mill, and "not making much headway," according to his daughter Josephine Trunkel.

While immigrants learned about the wonders of the new country from guidebooks such as Ole Rynning's *True Account of America*, American workers read advertisements or articles about "rich, fertile farms" on handbills or in their native-language newspapers.

Peter Jasicki read a Gates Land Company ad in Milwaukee and moved his family to Rusk County. At Weyerhauser, he became a land agent himself and recruited fellow Poles. The Cypreansen Brothers of Eau Claire hired Vincent Benesh to advertise their land holdings in Bohemian language newspapers. By 1905, more than 100 families had settled around Drywood. During this period, commercial dairy farming began to dominate area agriculture.

But while plots were affordable, "life was hard" recalled Mary Snedic, who grew up near Willard in the 1920s. "The land was full of stumps and stones. Everything had to be grubbed by hand or with horses." Esther Schrock found the same was true at Glen Flora. "It just rains rocks here," she said.

In the Bloomer area we were fortunate, we didn't have much rock.
When we moved to New Auburn, we learned what rock was.
  — *Don Moos, Chippewa-Barron County line, interviewed 2000*

Cooper family hauling in rocks from their fields to use in the foundation of their new barn, 1948, eastern Taylor County. Left to right: Lorence, Herbert, Inez (Robertson), and Wayne Cooper. Courtesy of Inez and Roger Robertson.

Time

It was a log house. My dad built on a dining room and a kitchen after they bought the place. The original building was all log. Cold, whoo, you don't know what cold weather is.

— Cleve Kirkham, Augusta, Eau Claire County, interviewed 1998.
His father bought the farm and farmhouse in 1905.

It was an old log house and … you had to go outside and climb up a ladder and then climb into the window to get in the upstairs. And that is what we lived in for the first two years.

— Ethel Heath, interviewed 2000, speaking of their first house she lived in
with her husband Tom, near Tony, Rusk County, in 1940.

In telling the tale of Chippewa Valley farming, telling the time is a complicated matter. Time spreads out like a ripple from the railroad tracks and rivers. Generally speaking, the farther a farm, a neighborhood, or a community was from a "hub" — a city such as Eau Claire or Menomonie, a travel route such as the Chippewa River or U.S. Highway 12 — the later any change might occur. These five families all settled a parcel; cleared it of trees, brush, rocks and stumps; built houses and barns; and started working the land: the Andersons, Chippewa County, 1857. The Solies, Dunn County, 1880. The Perovseks, Clark County, 1908. The Penskovers, Barron County, 1918. The Heaths, Rusk County, 1940. In some ways, each family shares one story. And in a way that is very important to this narrative of Chippewa Valley farm life, each of their stories takes place at the same time, the time that a farm began.

We started out from Eau Claire. A terrible storm came up with much lightning, thunder, rain, and darkness. The roads were poor and in some places there were no roads at all. We came to Philip Henrich's place, and there Big Gunder inquired, "How far is it to the end of the world?" Having secured the neccessary information, we drove on.

— Oline Arneson, writing in 1856

I remember that [father] picked my mother up and I with a horse and cutter and we stopped at the Bradford farm and had dinner, then he brought us up…. There was no roads in here and we had to go through the woods. And I can remember the horse fell through the [ice] when we crossed the creek.

— Carl Penskover, interviewed 1998, speaking of the day
his family moved to their farm near Rice Lake, March 9, 1918

Farm Life

I always kid my dad, tell him that they used to push me between cows in a stroller to make me milk… We raised our [own] kids in the barn. Two days after they were home from the hospital they were in a buggy in the barn doing chores with us.

— Jeff Pagenkopf, Elk Mound, interviewed 2000,
speaking of the 1950s and the 1970s

My dad said if you're a farmer there is three things: He says, you'll never go hungry. You'll never get rich. What else did he tell me? "You'll never run out of a job." And I said, "Why in the hell can't a farmer get rich as well as anybody else?" He said, "If they did, who would do the work?"

— Tom Heath, Tony, interviewed 2000,
recalling a conversation that probably took place in the 1940s

Arthur and Adelaide Kohlhepp, about 1925, Eau Claire County. Courtesy of Elmer and Margery Kohlhepp.

Top: Flailing beans inside house.
Chippewa County, Wisconsin.
September 1939. Photograph by
John Vachon. Library of Congress,
Prints & Photographs Division,
FSA-OWI Collection, fsa
8c36029u.

Bottom: Family at the Kysilko dining
table, June 1941. Courtesy of
Jeanne Kysilko Andre.

Managing Farm and Family

Especially when they used to milk by hand, she milked half and my dad milked half. Then when they got the milk machine my dad did all of that. My mother used to help in the barn quite a bit, like washing the milker and feeding calves. But my dad basically ran the farm and my mother took care of the house, garden, chickens, and raised the kids: my brother and I. I don't know — it worked for us.
— *Eugene Felix, Stanley, interviewed 2000, speaking of the late 1930s and early 1940s*

The farmhouse is the sphere where the labor of farm husbands, wives, and children intersect. In city life, many people have considered the household as the woman's sphere, a supporting role to the economic sustenance of the family, which happens somewhere else, in a mine or a factory or an office. But a dairy farm home *is* a business. If the barn is the factory floor, where the product is produced, then the farmhouse is the office, where the business is managed. For centuries, farm women have contributed directly in this sphere, producing food as fuel for the workers, economizing to balance the balance sheet, and often keeping the books. During the lumbering era, women were often the primary dairiers as well as the primary keepers of the home. By the middle of the twentieth century and after, men usually led the barn-and-field team, and women led the house team.

As farm women's roles changed later in the century, their self-perceptions grew more complicated. After a lifetime of baking her own bread, Irene Parker of Chippewa County was so embarrassed the first time she went to buy bread at the store, she almost didn't go through with it. And Inez Robertson acknolwedged the irony of never having qualified for Social Security because she did not work off the farm. Inez often worked more than four hours a day in the barn, raised eight children, and produced much of the family's food, clothing, and household items.

Any money that I've ever made… it doesn't go in my pocket and Don's money doesn't go in his pocket. Our money is directed at working together, through business or through family…
— *Ilene Moos, Chippewa-Barron County line, interviewed 2000*

The Home Economy

There was, officially at least, no "Wisconsin Farm Family of the Year" in 2001. That year, the University of Wisconsin–Madison's Center for Dairy Profitability cancelled the program, citing a lack of applicants. Organizer Bruce Jones, professor of agricultural and applied economics at UW–Madison, wrote in a letter to the award sponsors, "It is my guess dairy producers were reluctant to apply for this year's program because they felt their earnings in 2000 were not high enough to qualify for the award."

In real inflation-adjusted terms, milk prices were as low in 2000 as they had been in 1932, the bleakest year of the Great Depression. By 2004, milk prices had recovered significantly, but that suggested chiefly that prices were volatile.

Wisconsin dairying had also changed since the early twentieth century, when home production of food, clothing, and other necessities — and sharing work with neighbors — reduced expenses, while selling or trading farm produce from diversified operations brought cash to the farm. Everyone had a place in the family economy. Dan Emmerton said of his childhood near Colfax, "I never could understand why I always had to go home and feed them damn chickens, on Sunday, when it was my time off. Later I kinda figured out that that was part of the thing of knowing where the heck I was."

As the twentieth century had progressed, farm operating costs had gone up, and many expensive consumer items such as electrical appliances and automobiles had come to be seen as necessities. In 2002, on 40 percent of Wisconsin farms, at least one spouse worked off the farm 200 days or more. With work in the barn and fields still to do, there was less and less time for labor contributions to the household economy. It became easier, but more expensive, to buy the family's clothing instead of making it. It became easier to buy packaged food instead of making meals "from scratch," or for the spouse working off the farm to get take-out on the way home. Still, any evening may have found balance sheets, bills of sale, and breeding records spread out on the kitchen or dining room table after the dishes were cleared and done.

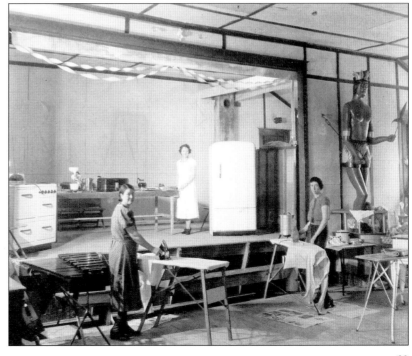

Ironing contest at the first annual meeting of the Chippewa Valley Electric Cooperative, held in the Holcombe town hall, February 5, 1937. Electric companies and cooperatives sponsored such events to promote "modern" appliances.

Making Do

Plastic braided rug, 2003, made by Inez Robertson from recycled plastic bags.

At one time, flour sacks were printed in patterns like you could buy, yard goods. And they'd use flour sacks if they didn't have enough money to buy yard goods at the local store. And some of them, you know, looked nice.

— *Phyllis Berg, Hale, interviewed 2000, speaking of the 1930s*

We would often trade feed sacks with other neighbors, or when we got feed, we'd try and get enough of the sacks alike ... we made our everyday sheets. Mama used to call them our twenty-dollar sheets because [it took four feed sacks to make a sheet, and] feed was about five dollars a bag.

— *Esther Schrock, Glen Flora, speaking of the late 1920s or the 1930s*

In their role as mechanics, farmers "have fixed what they could and used what they have," a survival mechanism they've often called "making do," and they taught their children this skill. In the house, homemakers have done the same, and taught their children. Darn a sock, use a Pepsi bottle as a calf's "baby bottle," make a braided rug from old woolens or a bunch of rags.

In the fields and barns, families have used their knowledge, skill and inventiveness to increase production and keep costs down. In the garden and house, farm families have done the same. Plant seeds and feed chickens, harvest and butcher, preserve and store, cook and eat.

America is going to be extremely in sad shape if we don't have more farmers and younger generations of people being able to have that experience of farm lifestyle ... or country lifestyle of being able to get along with what you have ... and not always having to be able to run down to the local store and buying it off the shelf. Sometimes you have to make it. Everything was made years ago right on the farm or local general store community, blacksmith ... and we are really in a sad shape of not being able to solve a lot of our simple problems.

— *Tim Dotseth, Menomonie, interviewed 2000*

Flour sack patterns. Inset: Margaret Russell wearing flour sack dress, 1925. Inset: Farm family with heating stove made from oil drum, Chippewa County, September 1939. Photograph by John Vachon. Library of Congress, Prints & Photographs Division, FSA-OWI Collection, fsa 8c36069.

Farm Fresh

A neighbor raised goats and they gave us several little kid goats and so we raised those and we butchered one for meat and the other we sent back to the neighbor. When those goats were small, we so enjoyed playing with them. We even let them in the house at my mother's permission, and they would run around the house.

— *Margery Kohlhepp, Eau Claire County, interviewed 1998, speaking about the 1930s*

During the Great Depression of the 1930s, Dorothy Bullis Carpenter remembers, the Bullis family always had enough: "You know, we grew the potatoes, we grew the vegetables, we grew the meat. So, we always had food when other people were maybe hard up for food."

Among the advantages of living on a farm: families have fresh food and they know where it comes from. They certainly raised a lot of it. There were two parents, four children, and the hired man on the Weinzirl family near Eau Galle during the 1930s. "So there was always seven of us to sit up to the table," said John Weinzirl. "[Mother] canned everything, from the meat to the berries. She put a lot of sauerkraut. I have real fond memories as a kid helping to make sauerkraut in the evening after the chores were done." Bob Donaldson's parents "raised everything." They only bought flour, sugar, and coffee.

Farm woman holding canned goods. Chippewa County, September 1939. Photograph by John Vachon. Library of Congress, Prints & Photographs Division, FSA-OWI Collection. Digital ID: fsa 8c36028.

Darlene Honadel recalled the fall sweet corn harvest on her family's farm near Augusta in the 1950s. "When we froze or canned corn ... that was a one-day thing. Dad would get the pick-up, we'd go down to the field where we always had the sweet corn planted ... and we would pick that pick-up load full of corn. And then you would husk, and as soon as somebody got some husked, then it went to the brushing and to the cutting so that you could keep going, and we would probably have 70 to 100 quarts of corn that we froze in a day."

Families made fun out of the work. Esther Bandli remembers preparing fifty pints of peas to freeze when her children were young. "Well, peas are pretty labor intensive. But we'd get 'em picked and we'd all sit on the porch in the evening shelling ... and have some home-made root beer floats."

The land also provided volunteer crops. Audrey Erickson's parents had apples, strawberries, raspberries, gooseberries, and currants. At "blueberry time" near Cleve Kirkham's Augusta home farm, he said, "We would take a team wagon, and would take along a couple cans of water for the horses and some hay and all the equipment to pick berries with and the whole family would go.... We would come home with pails and pails of them. After chores at night we would dump the whole lot of them right in the middle of the dining-room table and we would all get around the table and pick over blueberries. Mother could can them and make pies. Boy in the winter time they sure tasted good."

In 2000, some farms still had big gardens and raised a few animals for home use. Dan and Jan Weiss raised both "butcher-chickens" and laying hens, and they "butcher a cow whenever" on their 300-cow farm near Durand. Jan made her own spaghetti sauce and salsa and pickled beets and eggs. She canned "like crazy," she said. "I don't buy a lot of groceries out of the grocery stores like a lot of people do, just because of all of the things that I do can."

My dad, he liked lamb, so he'd always
butcher a lamb about August, you know.
But that's something, I don't know, they say
it's because I had pet lambs, but I just never cared for lamb.

— *Jim Solie, Menomonie, interviewed 1998*

Husking sweet corn for canning, Black family farm, Arkansaw area, Pepin County, about 1948. Clockwise from back left: Bernadine Black, JoAnn Fisher, Bill Black, Jim Fisher, Jean Black Kannel (standing), Jim Leffring, Eunice Black (back turned), Joan Liffring, and Janet Black. Courtesy of Jean Kannel.

Quantity Cooking

You were always washing dishes and cooking. And cooking and washing dishes. Someone who used to work for us told me a few years ago, he said, … 'You were always cooking, cooking, cooking.' I like to, so I was doing what I liked.

— *Margaret Kent, Rusk, interviewed 1998, speaking of the 1920s*

Inez Robertson of Sheldon cooked for her family of eight. "I had to bake bread three times a week, and I was kind of tired of always slicing the bread for everybody, so I started baking buns and everybody really liked the buns, especially warm with butter on and jelly… then I didn't always have to slice bread, so I'd bake 90 buns three times a week." With fuller houses during the holidays, women cooked even

33

more. And during the threshing era, they also prepared meals for visiting work crews of up to twenty hungry men. When the corn-shredding crew went from farm to farm, John Weinzirl said, "Gosh the ladies all tried to outdo each other with pies and all that stuff."

We always had lutefisk and lefse and those kinds of things, which I still don't eat.... Mother would always make blood bologna and stuff like that, which my dad just loved, and I never ate that, either.
— Bernice Sutliff, Menomonie, interviewed 1998, speaking of the 1940s

Neighboring

The neighbors would come here, and they'd go over to the neighbors and visit, and they'd sit and play cards and brew a pot of coffee. And I can remember them telling about this Mrs. Peterson over here. My folks were over there playing cards and, "Oh, it's time to make coffee," she said. And she'd put the coffee on the wooden cook stove ... and go back and play cards and that coffee would sit there and boil and boil and boil. So the coffee would be so thick....
— Byron Berg, interviewed 2000, speaking of the 1930s or early 1940s

Above: Playing cards, late 1940s, advertising Sunlite Dairy in Eau Claire.

Below: Stahlbusch family gathered around their kitchen table, 1949. Courtesy of Harold Stahlbusch.

When Margery Kohlhepp was a girl, whenever it was anyone's birthday or anniversary anywhere in the whole neighborhood, which was quite often of course, the adults would get together at someone's house, eat, and play cards well into the evening — as many as 20 or 25 people. "The nice thing about playing cards," she said, was "all the visiting you do while you play." The children would stay up until they were too tired to keep their eyes open, then they'd pile into the bedroom where all the coats were piled up on a bed and nestle down into the coats until their parents gathered them up.

Just as farm neighbors would borrow tools or machinery back and forth, they would also stop over at each other's houses any time of the day, especially in winter when farming didn't demand so many of their daylight hours. Summertime visiting might take a different form. Nathan Splett worked with his father in the fields on his parent's Barron County farm. Often his father ended up visiting with his neighors at the fenceline. "I used to really enjoy those conversations, and listening to my dad and the neighbor visit about what was happening." If need be, some neighbors had a kid-safe conversational filter. "It was a Norwegian neighborhood," said Eric Fossum, recalling his childhood near Rock Falls in the 1930s, "and they'd speak Norwegian all the time, especially if there was something that we weren't supposed to hear."

Neighbors also hosted get-togethers on the spur of the moment. Margaret Severson, who lived near Eau Claire, called up the three or four couples if she'd baked donuts that afternoon. It worked much the same in Esther Harriman's Augusta neighborhood. "The telephone would ring," she said. It would be the familiar voice of a neighbor on the party line, shared by several families: "'We're going to have a party at our house tonight,' and that was all they said. And then of course each one brought something, cake or sandwiches."

You never called up and said, "Could I see you at 1:00 on Monday," like you do today … because people, people just didn't do that. They'd just go to your house and rap on the door and you'd say, "Come in," and you'd put another plate on the table.
— Alyce Myers, Menomonie, interviewed 1998, speaking of the 1930s

The Party Line

Almost universally, rural telephone systems ran as "party lines" at one time. Several farms — generally eight or ten, but sometimes as many as thirty — were connected directly to each other on a single loop, and the farms on this loop shared access to a single switchboard connection in a nearby village or town. So, when a person rang one phone on the party line, she rang them all.

A crank on the side of the phone operated a magneto inside it, which generated electricity as long as the crank was being turned. Phone lines carried this electricity to the other phones on the loop and operated a clapper between two bells on each phone. This allowed a caller to ring everybody's phone in a pattern: crank a long time, and everyone hears a long ring; crank a short time, and everyone hears a short ring. Individual farms were identified by the ring code, which everyone on the party line, and the central operator, kept in mind. The Johnson farm might be two longs and a short, while the Peterson farm might be a long, two shorts, and a long.

The party-line system dominated rural telephone systems for decades. The Bell System introduced party lines in the late nineteenth century as a way to speed the spread of telephone service, with the hopes that people would quickly switch to the more-expensive private service. For a variety of reasons, the largest urban centers switched to private service decades before rural areas. But, still, in 1950, party

Telephone, 1908.

35

lines made up about 75 percent of all U.S. telephone lines. Some regions of the U.S. had party-line service in 2000 and after.

The system worked very well, except for a lack of privacy. *Long–two shorts–long* might signal that the call was meant for the Peterson farm, but the Johnsons could pick up if they wanted to. "Everybody on the line listened," said Merle Sjostrom. "That was just taken for granted." Her husband Edwin concurred. "If the phone ever rang at night, you could be sure there would be some broken toes, 'cause the neighbors would be trying to get their phone and see what was going on." But, Merle continued, "There were nice things about it. If there was something really happening, some event, or somebody needed help with something, the central would ring a bunch of little, short rings."

> *If someone had a fire, they went to their phone and they'd ring the crank seven times, that was the signal for a fire, and at that point, everybody on the line would run to their phone and lift their receiver to see where the fire was…. And then you made a dash for your car and went to the place and helped put out the fire because there were no rural fire fighting units at the time. So the neighbors really cooperated and helped each other.*
>
> *And soon after that, the women would say, "Well, we'd better get some lunch together and get over to that fire." They'd make sandwiches and coffee and get there and stay there for hours while the men were working to get the fire put out.*
> — *Margery Kohlhepp, Eau Claire County, interviewed 1998, speaking about the late 1940s*

Money Matters

When Jim Solie's parents wired their house and barn around 1940, they put one light and maybe a couple of outlets in each room. They put six 25-watt bulbs in their 80-foot-long barn. "They were very careful on what they bought." Alyce Myers remembers that the family didn't get to Menomonie very often, "and you never forgot anything that you needed, either, because you wouldn't get it until the next time you went to town. If Dad had any extra money, we'd get an ice cream cone. If he didn't, we didn't get any. It was a big deal to get an ice cream."

As a boy in the 1920s, trying to sleep upstairs, Harris Kahl could see stars through the roof of the house. Water would freeze on the kitchen floor when Audrey Erickson was a child in the 1930s; inside a neighbor's house, one of their children got frostbite. They learned to be thrifty.

Bringing Cash to the Farm

You think of anything you can do to generate some income. Well one year I know we raised potatoes and sold those door to door in Eau Claire. We would take these laying hens, what they called spent hens today, and we would butcher them and then take them into Eau Claire. We would throw them in a stock tank with ice to cool them down right away and take them into Eau Claire and maybe this little corner store would buy six of them, and another one would buy 12.

— La Verne Ausman, Elk Mound, interviewed 2000, speaking of the 1950s

Making a purchase at farmer's roadside stand, Eau Claire County, September 1939. Photograph by John Vachon. Library of Congress, Prints & Photographs Division, FSA-OWI Collection. Digital ID: fsa 8a04949u.

In the late 1800s, Russell Doane's grandparents raised chickens and hogs, butchered and dressed them on the farm, and hauled them into Eau Claire to sell to butcher shops or at the Haymarket, at the confluence of the Chippewa and Eau Claire Rivers. In the 1920s, John Weinzirl's dad butchered hogs and peddled them house-to-house, and in the 1930s and 1940s, Margery Kohlhepp's parents dressed turkeys and delivered them to Eau Claire houses at Thanksgiving and Christmas.

Farm families have employed many different strategies for bringing cash to the farm. Inez and Roger Robertson sometimes had to sell a cow to pay the dentist's bill for their children. Cash crops, timber, and the land itself provided liquid assets. Some families sold craftwork or set up roadside stands. The farm provided a small wealth of opportunities.

Egg and Butter Money

When Merle and Edwin Sjostrom of Maiden Rock were first married in the 1950s, they raised pigs and cows — and chickens. "I remember taking, probably a half a case of eggs downtown to get groceries," said Merle. "Grocery stores took eggs in trade, and we'd always take them down to the grocery store."

These were not small flocks. Around 1940, Myron Wathke's family had 375 laying hens. One of the family's egg cases held 12 dozen and another held 30 dozen. Myron cleaned the eggs and packed the cartons daily. "That was a job I dreaded. Your day's work was done and then you would have to pack those stupid eggs [to sell at] a couple of places in Fall Creek, had one in Foster."

The cleaning of the eggs, especially, was tedious, a job that many children and women disliked. But it was important to the home economy. The milk check and the seasonal sale of the field harvest provided bigger checks. Egg money provided store-bought items, including foodstuffs, and maybe a little cash flow.

A lot of your groceries at that time came from the sale of eggs. Every farmer had a hundred hens or so, and you'd take that crate of eggs to town and sell it, and buy groceries back. And many a times your upstairs looked like a storehouse because a lot of places would not give you cash. They would give you groceries in trade, so you would have canned goods — corn, peas, beans, that sort of stuff — shelf upon shelf.
— Don Foiles, Cadott, interviewed 1998, speaking about the 1940s

Woven-wire basket used by three generations of the Schumacher family to gather eggs. First used between 1900 and 1920, Elk Mound.

Off-Farm Work

From the 1950s through the 1980s, Inez and Roger Robertson raised eight children on their Rusk County farm. To bring in extra cash, Roger worked at an implement dealership, cement plant, and grocery store. While Roger worked off the farm, Inez managed the barn and house chores. At the same time, she made and sold braided rag rugs to help stretch the farm family budget a little further.

Working off the farm was a strategy employed by almost every generation. Audrey Erickson's grandfather came to Menomonie from Sweden, even though as the eldest son he would have inherited his parents' farm in the Old Country. While farming near Menomonie, he worked at the Rice Lake logging camps and as a "track walker" for the railroad.

As more women joined the off-farm workforce, the transition could be emotional as well as financial. Both Don and Ilene Moos agreed that "the day she went to work" was the most difficult time they could remember on their farm. "I was raised in the fifties," said Don. "When I was raised, the husband did the work and the wife stayed home and took care of the kids, you know, and cooked the meals and everything like that. And the day [Ilene] went to work felt like I had failed, I guess, you know.... And it took me a while to get over that.... I know afterwards that she did what she knew she had to do, to do the best thing for our family.... So, it probably brought us closer in the long run."

In 2000, Gary Evans was a farm management instructor at the Chippewa Valley Technical College in Eau Claire. Among his married students, only 10 percent of the wives worked on the farm; the other 90 percent worked "in town." Many spouses work as much for the health benefits as the income. Otherwise insurance could be a huge drain. In 2000, the Siverlings of Bloomer paid $400 a month ($4,800 a year) for insurance for their family of six — with a $3,000 deductible. "If something major happens, yeah, we're covered," said Steve Siverling, "but otherwise we have no health care to speak of."

> We used to have traveling salesmen through the area, it wouldn't do them any good now because they wouldn't find any of the women at home.
>
> — Ethel Heath, Tony, interviewed 2000

Family Partnerships

Most area farm couples made financial decisions as a team, which sometimes meant agreeing on decisions and sometimes meant delegating decisions to each other. "My mother was more of the type to live as you are and don't spend a penny," said Don Foiles. "She was very, very close with the money. And my dad was, was more of a maverick-type farmer, you might call him. He wasn't scared to take a chance on a crop, and of course that's what got you ahead, too.... But basically the farming was his decisions, but the household was more mother's."

On the Siverling's Bloomer farm in 2000, there was a meeting of the minds: "[Kim's] the book-keeper," said Steve. "And if I shot an idea that I want to do something in a certain way, as far as spending money, then we, then it's a two-person deal on this. I don't go out and buy a piece of machinery unless I run it by Kim and say, 'Well, you know, I think we need this and I think it will pay for itself in such and such a time', and I got to do a sales pitch on her, you know, to go to work and do it. As far as the crop and everything, that's all my own decisions as far as what's going to get planted where, when we do that kind of stuff. The same way with the cattle as far as the breeding program or any of that.... Then again, on the other thing, when it comes to the family, then it's all Kim [but] she runs it by, we talk back and forth."

Some farmers, like some people in any business, have gotten themselves into trouble. Fall Creek veterinarian Dennis Stuttgen saw it on his rounds. "You go to some farmer ... and they got this big plush place, they got fancy 4-wheelers in front of the house, and a mobile home, the boats, and everything else, you go to the barn and the cows are ... not just dirty, [but also] just crap, no good. This guy is not going to stay in farming. Is he a success? Well everybody who drives by thinks he is. Another place you go by and it may look plain and may look small, but he has got more than enough money to live on, to send his kids to any school they want to go to."

The Five Ds

Those boys who came and worked [on our farm], they were kind of like members of our family, too. They were neighbors.... Their parents had a nice farm, but things happened. Their father and mother both were addicted to alcohol. And they had about five or six children. And as time went on it destroyed their family life. And those two boys ... usually they would just eat dinner and supper [at our place].... It was kind of sad what happened because the family broke up, moved away, got scattered.

— Audrey Erickson, Menomonie, interviewed 1999, talking about the mid-1930s

Farming is full of "what if's." What if we bred this cow to that bull, since he's known to throw calves will grow up to be fine milkers? What if we plant soybeans on the south forty this year, since last year's soybean price was better than for corn? At a "Heart of the Farm" conference in Eau Claire in March 2003, Joy Kirkpatrick, Richland County Extension dairy and livestock agent, asked 60 farm women in attendace to ask themselves "what if" questions they don't want to ask. What would happen to the farm business if a tornado hit? What if Mom and Dad got a divorce? What if one of the siblings wanted out of the farm?

At the conference, Kirkpatrick spoke about "Five Ds": death, disability, disaster, divorce, and disagreements, and said that insurance, estate planning, and shared management have helped families survive such setbacks. University of Minnesota family social scientist Pauline Boss told *The Country Today* in 2001 that "ambiguity, more than the event of loss, can immobilize and depress.... It's not knowing what is happening or what might happen, not knowing what you are doing wrong, what you can do to fix the situation.... These feelings all lead to stress that can reach dangerously high levels and can result in too much drinking, verbal and physical abuse of loved ones, and even suicide." The University of Minnesota Extension Service offers a publication on the uncertainties of farm life called "Losing a Way of Life? Ambiguous Loss in Farm Families."

In her discussion, Boss added four more Ds to Kirkpatrick's list: depression, drunkenness, drug use, and domestic violence. Statistics on such farm problems are often difficult to separate from rural problems in general. Against popular perception, the *rural* population of the Midwest generally held steady from 1940 to 1990. But the *farm* population fell dramatically. Consequently, farm families, who made up more than half the rural population in 1940, made up about 10 percent of the rural population 50 years later.

Recently, several researchers have paid significant attention to such problems among farm families as a subgroup of rural families. In 2000, University of Wisconsin–Madison professor Roger Williams blamed "chronic prolonged stress" for depression among farmers. He told listeners at the Wisconsin Farm Health Summit that at one time farm families were "incredibly resiliant," but after 1980, a combination of stressors eroded that resilience.

According to the Centers for Disease Control in Atlanta, the Wisconsin Farmers Assistance Hotline received an average of nearly 1,000 calls per month in 1994. Farmers called for a variety of reasons, but, "increasingly, the hotline [was] fielding calls about serious emotional problems: depression, withdrawal, alcohol abuse, domestic violence, and suicidal intentions."

As a boy at the beginning of the twentieth century, Melvin Christopher was in a party searching for a missing neighborhood hired man. He vividly remembered finding the man hanging from a tree. Tough times at the end of that century didn't make things any easier for a few. In 1999 and 2000, according to Laurie Woods, MS, and other researchers at the Firearm Injury Center, Medical College of Wisconsin, "Wisconsin's farmer suicide rate (18.2/100,000) was higher than overall rural (11.4) and Wisconsin statewide (11.3). In 2000, 12 of 14 farmer victims [86 percent] were dairy farmers although only 27 percent of farms are dairy." The researchers concluded, "Increased suicide risk among Wisconsin farmers, particularly dairy farmers, may be related to farming policies, physical illness, financial problems, and access to firearms."

In a 2000 story on rural domestic violence in Wisconsin and Minnesota, the Associated Press noted that "Isolation, fear of gossip, and worries over who will feed the livestock are all obstacles for rural women trying to flee domestic violence." The same year, a study by the *St. Paul Pioneer Press* found that about one in four domestic killings in Wisconsin took place in its most rural counties. The response is also complicated by rural isolation. "Because of the rural settings, we don't have the immediate response times," Polk County Sheriff Ann Wade said in 2000. "It can sometimes take us 10, 15, 20 minutes to respond to a domestic violence call, compared to a metropolitan area, where the response is probably within five minutes."

Keeping the Farm

You can't pass the farm down; you have to sell it to your son or your daughter. And that's what's killing them. Because [the farm has] to be appraised, and they have to pay the appraisal price. And the kids can't afford that. You're starting $80-$100,000, more than that even, in debt. That's just for the bare land, then, you got to have your cattle, your machinery. So, by the time you got it all done you probably got a quarter of a million dollars you're in debt. And you haven't even started to farm yet.
— *Jeanette Stabenow, near Eau Claire, interviewed 1998*

In colonial New England villages, householders were given village lots and some pasture land. How much land depended on family size and social status. A merchant with a big brood might get 200 acres, but even an unmarried, unimportant fellow might get ten. And it was his to keep and pass down. However, large American families, and long lives in the land of plenty, caused the population to explode — and the demand for land with it, as children grew into parents. News about the abundance of America brought more immigrants. Most of the good land near the good markets was taken within a generation or two. To add to this, parents were living beyond the time their children started families and needed land of their own, which wasn't so much the case in past centuries in Europe. Succeeding generations moved west and west again. Still, people had the right to pass land to their heirs if the situation suited.

Some people have argued that the inheritance tax, or "death tax," took away that right in the late twentieth century. Until 2003, the tax could take up to 55 percent of a decedent's estate. Assets of up to $600,000 per person were exempt, but farm land, with machinery and livestock, could quickly pass that mark, making farm families fear the death tax. However, in mitigation, family-owned businesses and small farms received special treatment under the tax, including large deductions when the business or farm represented at least 50 percent of the estate.

It's a complex issue, with family farmers on both sides. Some argue that unless the temporary diminishment of the tax made law during the George W. Bush administration became permanent, it would foreclose the birthright of their children. Others argue that the rollback was made with large, agribusiness farms in mind and would end up affecting less than 2 percent of estates.

Some families don't wait for death and inheritance. Children buy farms from their parents when the parents retire. But farmers argue that while once they could have given their children a deal, this became impossible in recent decades. In turn, this engenders another difficulty, attorney Stuart Urban told Janelle Thomas, a correspondent for *The Country Today*, in 2003. While farms have a lot of value, they often don't produce enough income to support payments on a mortgage at a current full market value. The elder generation paid less for the farm when they bought it years ago and consequently had lower mortgage, or they've long since had that mortgage satisfied.

To be the last generation to farm a particular piece of land can be a daunting decision. Five generations of the Gilles family have lived on land near Cadott. It was a dairy farm, but beginning in 1945, the Gilleses set up a series of inventive enterprises to keep the land supporting the family, including custom meat processor, mink ranch, Christmas tree farm, and recreational facility.

Other farms sell at auction. Unlike an urban center, where one family enterprise might benefit from the success of the business next door, members of farming townships can find themselves in the paradoxical position of sometimes benefiting from the fact their friends and neighbors have quit farming. John Weinzirl, speaking in 2000 of his farm near Eau Galle, was not unaware of this paradox: "The farm that we have now had seven families making a living on it at one time."

In the best cases, auctions benefited farm families handsomely, at the time the operators retired or died — these sales could be pageants of a successful career or life. And other farmers benefited as well: As Phyllis Berg of Hale noted in 2000, "Machinery is extremely expensive" and many farmers "can't buy new machinery, so they go to auctions." But especially during the Great Depression of the 1930s, and the farm crisis of the 1980s, auctions also symbolized the unhappy end of an operation.

While much about the issue is both shaded and emotional, two things seemed clear to farm families in the 1990s and 2000s: that as more farm families quit farming and their farms passed out of the family, the farms that were left were getting larger; and that farms of any size, but particularly large farms, have gotten harder to pass down to the next generation.

> *The yard was muddy. A late winter snowstorm had blanketed the farmyard with six inches of wet, sloppy snow. It seemed only fitting. The farm looked old, gray, weathered....*
>
> *It had taken Dad more than 35 years to accumulate his animals, machinery, and tools, and in less than eight hours everything would be gone. I didn't remember his tools looking so rusted before. Was that the "good" Leland tractor? When did it lose its bright blue paint?....*
>
> *Dad looks tired. The farm accident last summer had taken a lot out of him. We almost lost him that day. The creditors are calling.... Dad and Mom are making the right decision. We're all here. All seven children, their spouses and grandchildren have come from throughout the state.*
>
> *The auctioneer is ready to begin.*
>
> *— Cynthia Hofacker, Eau Claire, from an essay published in the* Eau Claire Leader-Telegram, *December 31, 1999*

Barn raising at John and Barbara Schuebel's farm near Boyd, Chippewa County, 1903. Courtesy of Rita Schuebel and Stanley Area Historical Society.

Milking Night and Morning

By the 1890s, as wheat production moved to the Plains states, dairy became the focus of farming in the Chippewa Valley, though most farms were still diverse operations. Crops used to feed dairy cattle were well suited to Wisconsin's harsh climate. They also replenished the soil, which had finally been depleted by heavy wheat production. The dairy industry in Wisconsin supported a large number of small dairy farms and encouraged the stability of the small towns that surrounded them. Through the efforts of government, universities, and farmers themselves, dairy became both a science and a business, not to mention an art. And, in a dairy business, the production plant is the barn.

Mother and daughter, Marie and Rose Kubista, preparing to strain fresh milk, 1938, east of Shell Lake, Washburn County. Courtesy of Rose Kubista Tomesh.

Making a Family Business

Not much time to play. It was mostly work. [Mother and father] split wood and after school we had to carry those split chunks — maybe ten or twenty armloads — enough to last the whole day, because it was used for cooking, even in summer. And we had a big reservoir beside the wood stove, and — me and my sisters — we had to fill it with water every day after school so there was warm water in there. Eight years old I started to milk cows by hand. I milked four or five cows every day — and my brothers and sisters did, too. We had 16 head.

  — *Josephine Trunkel, Willard, interviewed 2001, speaking about the 1920s.*

In the 1920s, just before electricity came to the farm outside Prairie Farm, Harris Kahl and his mother and father milked 29 cows by hand, "night and morning." By mid-century, men and older boys

Vern Bullis carrying the basic equipment needed to milk by hand: bucket and stool, about 1920. Courtesy of Dorothy Bullis Carpenter.

did most of the milking on many farms, but other family members took over during busy times of the year or when he was away from the farm. Women often cleaned the pails, bottles, milk strainer, and the many parts of the cream separator.

Dairy production assured farmers more predictable profits than other cash crops such as wheat or tobacco. The steady income allowed families to buy their farms rather than rent, and in turn encouraged rural communities to flourish. Farmers organized to get better transportation rates from the railroads and better prices for their products, and they made improvements in their farms and herds that increased production, made milk safer, and kept it fresh longer.

Marketing Milk

They used to milk the cows once a day. And they separated the milk in the house. And then the creamery at Downsville picked up the cream three or four times a week.

— *Clarence Werner, Weston, interviewed 1998,*
speaking of his parents' operation in the 1920s

When she was a girl in the late 1920s, Dottie Carpenter and her two brothers rode with their father as he took his milk to the dairy. They turned onto the road from their farm on the land where Eau Claire's Oakwood Mall was built years later. They bounced along the road, "our feet hanging off the back of the truck," jostling each other, but making sure to hang on. They waved to people they knew as they passed. At the dairy, once the milk had been weighed and paid for, they each got a free ice cream cone. "That was our biggest thing for the day."

Wisconsin already had a quarter-million dairy cows in the late 1860s and produced three million pounds of cheese. The great quantity of dairy products forced prices down in nearby markets. Farmers needed to sell milk over a larger area. Several changes helped this happen: Railroads made it practical to move perishable liquid milk beyond local markets, scientific advances speeded the grading of milk at markets, and organizations such as the Wisconsin Dairymen's Association, founded in 1872, had the resources to market state dairy products over a wider area.

By 1912, Wisconsin had 1.5 million dairy cows, made over 12 million pounds of cheese, and had surpassed New York to become the nation's top cheese-producing state. But the state was just at the beginning of its dominance. By 1935, one county (Clark) made 19 million pounds of one kind of cheese (cheddar). At that date, Wisconsin had 3.2 million dairy cows and 2.9 million people.

"Make the Good Cow Better"

If the boy comes from the farm and learns at the university how to make that farm more useful, and from the scientific methods which he acquires develops more orderly habits of life; if he receives some inspiration which leads to progress or some ideals which lead to good citizenship; is it not worth while?
— Charles McCarthy, The Wisconsin Idea, *1912*

Diploma awarded to Curtis Yule upon completion of his Bachelor of Science degree in Agriculture, University of Wisconsin (Madison), 1914.

In 1885, William Dempster Hoard of Fort Atkinson launched *Hoard's Dairyman* magazine, which was still published in Wisconsin in 2004. Through his magazine, Hoard advocated using alfalfa as a dairy feed, tracking milk production per cow, and using silage as feed. He bought a Jefferson County farm as a laboratory to test the ideas he was promoting. Hoard was elected Wisconsin's governor in 1888.

Meanwhile, the University of Wisconsin (Madison) made and promoted numerous breakthroughs in dairy science. These breakthroughs led to selective breeding in cattle, the eradication of milk-borne tuberculosis, dramatic advances in seed strains and field yields, and the education of Wisconsin farmers through its "Short Course."

Many of these improvements required farmers to keep detailed herd, milk, and business records. Not all dairy farmers adopted these new methods, but those who did made Wisconsin into America's Dairyland.

The Regents of
The University of Wisconsin
on the nomination of the Faculty
have conferred upon
Curtis L. Yule
the Degree of
Bachelor of Science (Agriculture)
together with all the Honors, Rights and Privileges belonging to that Degree. In Witness Whereof, this Diploma is granted bearing the seal of the University and the signatures of the President of the Regents and the President of the University.

President of the Regents. President of the University.

Given at Madison in the State of Wisconsin, this seventeenth day of June in the year of our Lord nineteen hundred and fourteen and of the University the sixty-fourth.

Bigger and Brighter

At mid-century, Wisconsin dairy farmers still subscribed to an image of themselves as independent farmers, whose communal labor formed the backbone of the nation. However, advances in science, technology, and government regulation were radically changing the nature of farming.

Electricity and automation in the barn saved labor. Before Cleve Kirkham got a milking machine, he was milking fifteen cows by hand on his farm near Augusta, and "it was quite a job." Public-health ordinances and mechanized handling systems made milk more sanitary. Government programs sustained domestic prices and discouraged imports. And, the merger of local cooperatives into regional and state-wide organizations tried to keep farmers from losing economic power in an increasingly complex market.

While farm incomes rose substantially during World War II and after, men and women were spending less time in the barn and fields, and more time maintaining production records and keeping the books. On Tom and Ethel Heath's farm near Tony, Ethel said, "He'd run things out in the barn and I'd take care of the house." But taking "care of the house" included getting the farm "ready for the tax man and all that kind of stuff" and keeping the "buying and selling" records.

Surge-brand milking machine, about 1960, used on Frank Meinen's farm, Chippewa County.

The Latest Equipment

When I was very small, we didn't even have a barn cleaner. We basically had to do everything with a pitchfork … And I can just barely remember this, when the snow got to be three or four feet deep in the fields, they would hook up two work horses to the manure spreader instead of taking the tractor out.

— *Bruce Donaldson, Eau Claire, interviewed 2000, speaking about the 1940s*

When La Verne Ausman was a boy in the 1940s, his family already had a milking machine. By 1955, after he left home and was farming with his wife Bev, he had a pipeline system, which took milk directly from the cow's udder into a bulk tank. No more cows kicking buckets, no more 10-gallon milk cans to lift. But Eugene Felix didn't get a pipeline system until the 1970s, and Bob Donaldson, who milked cows into the 1990s, never got one.

In the 1950s, electric water pumps, barn cleaners and silo fillers lightened daily chores,

Milking with a machine run by an electric vacuum pump, late 1940s. Courtesy of Chippewa Valley Electric Co-op.

but, added to big-ticket field machines, such expenses forced many smaller farmers out of business. The survivors found that new equipment could buy both ease and extra income from improved productivity.

Power milking machines saved labor, and fewer family members were needed in the barn. But power field equipment also meant that women and older children could share in heavy crop work. Wisconsin surveys in both the 1960s and 1980s found a traditional division of labor (field work for the husbands, home economy work for the wives, some shared barn chores and household maintenance). However, the later study showed that women's work was increasingly diverse (greater numbers driving tractors, raising livestock, managing business aspects) while men were no more likely to take on "household" tasks.

Farmer demonstrating the use of an electric silage unloader, early 1950s. Courtesy of Chippewa Valley Electric Co-op.

My dad would get up at five-thirty every morning and they would be milking at six o'clock, and we had breakfast at seven.... This gas engine that ran the milking machine ... made a ton of noise.... And when that stopped, we flew like crazy to get the potatoes done frying at supper time, or, in the morning, to make sure that breakfast was going to be on in a hurry, because we could tell when the milking machine stopped just how much time we had.
— *Dorothy Bullis Carpenter, Eau Claire, interviewed 2000*

Managing the Herd

You have a lot of things you have to be concerned about, you have diseases you have to contend with so you have to make sure your vaccination program is correct, and you have other [infections] such as mastitis that a farmer always has to be concerned about. You have to sometimes resort to antibiotics and you have to know which ones work and which don't and what the withdrawal period is to keep the milk safe and that is just in the dairy end of it.

You also have the breeding program you have to work with.

Then in the machinery end of it you have maintenance in there and you have to know as far as what type of equipment that is going to work in your soils and so forth.

So a lot of it is experimentation too on your own farm — what works for you may not work for another individual; it varies from farm to farm.
— *Alvin Kohlhepp, Eau Claire, interviewed around 1998*

Each dairy farmer is a herdsperson, veterinarian, mechanic, accountant, marketer, and CEO. Dairy farmers are also matchmakers: Pairing a specific cow with a specific bull can produce a calf better than either parent. And, since cows milk for a period after they give birth or "freshen," having cows give birth at more than one time during the year is essential to a steady supply of milk. Many farmers breed to have some cows freshen in the spring and others in the fall. They also need to detect, and record, the days each cow is fertile. Dairy farmers might use a breeding wheel, record book, or computer software. They are also attuned to the physiological signs cows give when they're ovulating or "in heat."

Fall Creek veterinarian Dennis Stuttgen examined the herd of a neighbor who swore he hated cattle. Stuttgen asked, "You hate cows? It all looks so good in there; the barn is clean, and all that kind of stuff. If you hate cows, how come you take such good care of them?" The farmer replied, "They make me happy." Other farmers, he said, professed to love their cows but had cows "lying around in different stages of disease and recumbence," because they said they "don't have time to take care of them."

"Why don't they have time to take care of their cows?" Stuttgen asked. He gave an example. "Because they haven't bothered to fix the water cups and so they are carrying water ... instead of fixing the water cups so [the cows] could water themselves. Poor management, poor priorities, things like this."

Rules & Regulations

For years, the Hayden family carried heavy milk cans from the barn to a detached milk house to cool the cans in a water tank. In the early 1970s, regulations forced George Hayden's father to install a bulk tank in the milk house, since the creamery could no longer accept milk in cans. He was upset about the change, but decided to let the creamery finance the pricey purchase "so if we go belly up they can have it back!"

U.S. regulation and inspection of dairy premises date to before the Civil War, and have come to fill reams. Each batch of milk is tested before it leaves the farm, and inspectors routinely visit farms to check for sanitation and procedural violations.

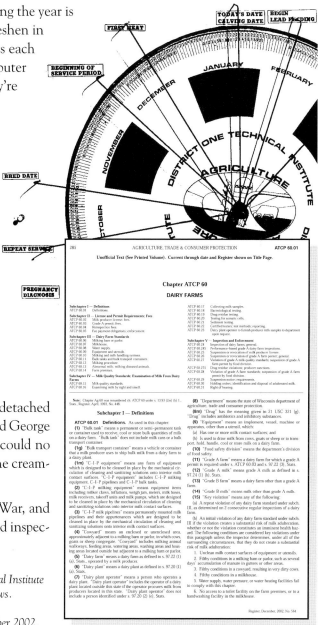

Top: Detail of a breeding wheel built by Loyal Pederson in a farm class he took at District One Technical Institute (now Chippewa Valley Technical College), 1970s. The wheel tracks gestational dates for individual cows.

Bottom: Page 285 from Chapter ATCP 60, "Dairy Farms," Wisconsin Administrative Code, December 2002.

51

Farmers have often lobbied for such regulation. One reason: Careless practices on one farm can ruin milk from neighboring farms once it's mixed in the milk truck or at the creamery. They also want the public to be confident of the safety of their product. However, farmers have historically felt resistance to government interference, and as a result, seemingly arbitrary rules cause tensions.

I read all the rules once: A milking stool cannot be padded. Whoever made the rules, I should be able to make their rules, too — that they can't have a padded chair behind their desk in their office.

— George Hayden, Pepin County, interviewed in 2000

Pest Control

Rats, mice, birds, flies, cattle grubs, lice, mange mites, and worms all affect the dairy industry. Flies expose cattle to micro-organisms and cause general stress in milk cows, which lowers their production. Rodents spread disease and spoil feed supplies. Gains in the amount and wholesomeness of milk have often come through control of these pests.

But poisons and pesticides can be toxic to people and to the natural enemies of the pests in question. Pesticides and antibiotics used poorly can lead to the development of resistant strains of pests, which are then harder to kill.

Farmers often gain great benefits from simple, sensible farming practices, such as storing feed in rodent-proof and bird-proof containers, keeping the barn and milking parlor clean, and cutting weeds in areas near the barn. Randy Wilson of Cadott attacked the rodent problem in their dairy barn with a method dating back to ancient Egypt; they got a half-dozen barn cats from a friend who had too many.

Top: DDT bottle, 1950s, distributed by Exito Laboratories, Eau Claire, for use in the home.

Bottom: Ten-year-old Luke Anderson spreading lime. Lime helps absorb moisture and prevent the spread of bacteria. Photographer: Dan Reiland. Courtesy of Eau Claire Leader-Telegram.

When DDT first came out everybody thought that this was great because you take some DDT and you spray it in that dairy barn and the flies were gone. And we were assured that it was safe and it was not a problem, et cetera. And it took, what, 25-30 years before we discovered the impact which that had on wildlife and the way in which it accumulated up the food chain.

— Fred Kirschenmann, sustainable agriculture advocate, Wisconsin Public Television, November 13, 2001

Hired Help

One time we went to Waukesha, Wisconsin, all five of us in our car, to a Brown Swiss cattle sale, where my father bought some new breeding stock. And that was a red-letter day — two days, I think — because in those days you just didn't travel very far, at least farm families didn't. But I guess we trusted the hired men at that time to take care of the farm and so we could go do that.
— Margery Kohlhepp, Eau Claire County, interviewed 1998, speaking of the 1930s

Near the end of the Great Depression, a young man walked onto Art and Anna Donaldson's farm. He came from South Dakota and was looking for a job. The Donaldsons gave him one. He stayed and worked on the farm thirty years until his death.

In 2002, Jeff and Marie Pagenkopf's Sandy Acres Dairy had 11 employees. The workers punched time clocks, got paid attendance bonuses, and received time-and-a-half on holidays and double-time on Thanksgiving, Christmas, and Easter. Only one lived on the premises.

In earlier decades, hired hands often lived on the farm. In the 1950s, the hired man lived upstairs at the Kent farm near Rusk. In the late 1930s, Bill Curry was the hired man on the Kopp's Dunn County farm. He slept with the boys in the attic above the kitchen, and the girls and their parents slept in the brooder house, while their new house was being built. The Morrow farm and the Sutliff farm both had separate houses for their hired help.

Other families hired help only when they needed it. In the late 1940s the Bergs, and in the 1980s the Siverlings, both hired high school boys to help with summer field work.

These big farms, they are well run, efficient, they are able to hire labor easier, because instead of having a hired man that works seven days a week, 15 hours a day, you get a person in that milks cows for eight hours and then leaves. You get another person comes in and feeds cows for eight hours and then leaves. So I understand why they are efficient and gaining in popularity.
— Eugene Felix, Stanley, interviewed 2000

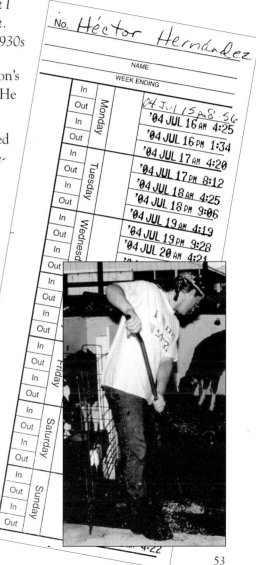

Hector Hernandez's time card, D&D Hawkins Farm, Chippewa Falls. In 2004, the farm had eleven workers from Mexico. Courtesy of Sara Hawkins.

Inset: Andy Rasmussen, Marie Pagenkopf's brother and paid employee at Sandy Acres Dairy near Elk Mound, 2000.

53

The Cost of Milk

In the 1940s, Elk Mound's high school band director was also one of three farmers within the village limits. His wife cleaned milk bottles, he filled them, and village residents came to their milk house and bought raw milk.

In most contemporary parlor-style dairy operations, after the cows are milked by machine in a milking parlor, the milk goes from the cow's udder through glass pipes to the milk house, a separate and meticulously-kept building. The milk travels into a stainless steel refrigerator tank that quickly cools the milk to just below 4°c. Milk is kept at that temperature until it is removed by the milk truck, which takes it to a dairy processing plant. All of the milk-handling equipment is made with convenient and thorough cleaning in mind.

In 2000, milking took up about one-half of the labor requirement on a modern dairy farm, and the milking facilities can make up half or more of the total capital investment in a dairy. Facilities and labor for milking can make up as much as 30 percent of what it costs farmers to produce milk.

"To-do" list from a day in the life of George Hayden, Pepin County, 2000. Written on a "dairy towel." Dairy towels, which are the size of commercial washroom paper towels, are typically made of bleached paper (as this one is), and are manufactured to be used in cleaning cows' udders.

Prices and Protests

We'll solve our problems with bayonets, and I don't mean maybe.
— Arnold Gilberts, Dunn County, September 1932. Gilberts was president of the Wisconsin chapter of the Farmer's Holiday Association

Halloween night 1933, Jim Roycraft's barn was burning. Fall Creek's fire department responded to the Roycraft farm in southern Chippewa County, but its tanker ran out of water as the fire raged. Since a milk strike was on, nearby farmers brought their milk to pour on the fire and help save his other buildings.

That very day was the climax of more than a year of passionate, sometimes violent, farm protests in Wisconsin. The Farmer's Holiday Association, which claimed 130,000 members out of Wisconsin's 180,000 farmers, had called a milk strike — during which farmers would dump their milk rather than take the low price they were being offered, as low as 60 cents per 100 pounds of milk in parts of the state. On Halloween, the much smaller but more vocal Milk Pool joined the strike.

Crowd gathered for a milk strike at Stanley, about 1932. Courtesy of the Wisconsin Historical Society. WHi 20483.

At least seven cheese factories were bombed and a farmer near Madison was shot to death. Two more Wisconsin farmers were killed and others injured in vehicle accidents as milk-hauling trucks were forced off the roads or overturned. The Farm Holiday ended its embargo in Wisconsin a few days later when five Midwestern governors met to form a plan "to save the farmer," and as quickly as the seething, year-long milk strike fire had flared, it flamed out.

Milk producers struck again on Independence Day and Labor Day in 2000, as milk prices hit lows not seen since 1979. During the Labor Day protest, farmers in 22 states dumped as much as 25 million gallons of milk. Bloomer dairy farmer Steve Siverling quietly spread his milk on his hay fields rather than pumping it into the milk hauling truck.

Right now we don't have a voice because we don't have big voting numbers. We need to do something to change that, and hopefully this [milk dumping] is a good first step.
— Ilene Moos, Bloomer, September 2000

Protests by Wisconsin Farmers Union members at the state capitol building in Madison, 2000. Courtesy of Wisconsin Farmers Union.

The Freedom to Farm

When Congress passed the Federal Agricultural Improvement and Reform (FAIR) Act of 1996, its chief goals were to end the Federal government's "acreage reduction program" — which gave farmers incentives not to plant crops the government thought would shortly be in oversupply — and to eliminate "target price/deficiency payments," a kind of subsidy paid to farmers when the market price of a farm commodity was lower then the government-established "target price." Supporters of the first goal argued that farmers, like other business people, could best determine what products to produce for the market; supporters of the second goal argued that a growing export market would send prices higher and eliminate the need for any kind of price support. These facets of the bill led people to call it, unofficially, the Freedom to Farm Act.

Unfortunately, a financial crisis swept through Asia beginning in 1997 and the Russian economy imploded in 1998. These events combined to force a dramatic decline in demand for agricultural exports from the United States. Then, in 1999, Brazil devalued its currency by 40 percent, making its agricultural commodities much more attractive on the world market. Exports of ice cream, dried whey, lactose, and whey protein, which had been expanding rapidly, began to decline, and the price for USDA Class III milk (used to produce cheese) dropped almost 40 percent in the few months between mid-1998 and early 1999 to around $10 per hundred pounds. In inflation-adjusted — "real" — dollars, milk prices hadn't been that low since 1932, the depth of the Great Depression.

The situation was made even worse for Wisconsin dairy farmers, because one small component passed with the massive 1996 farm bill was a provision that allowed six eastern states to form the Northeast Interstate Dairy Compact, a cartel that blocked the importation of Midwestern milk into its popular markets. After critics noted that it hurt consumers, fostered overproduction, and resulted in depressed prices elsewhere, the compact was shelved in 2001.

When Wisconsin dairy farmers had protested the 1996 farm bill, some observers criticized them as simply complaining about "the end of subsidies." But as the farm recession worsened in 2000, the federal government was forced to make billions in emergency payments to many kinds of farmers, including dairy farmers. The "Freedom to Farm" bill was heavily revised by a 2002 farm bill.

> *Despite the discrimination against dairy farmers in Wisconsin under the Federal Dairy policy known as the Eau Claire Rule, the 1996 Farm Bill provided the final nail in the coffin for many Wisconsin dairy farmers when it created and authorized for three years, the existence of the Northeast Interstate Dairy Compact.*
>
> — U.S. Senator Russ Feingold (D-Wisconsin), *testifying before the House Judiciary Committee, Subcommittee on Commercial and Administrative Law, June 17, 1999*

The Eau Claire Rule

U.S. milk prices have been more heavily regulated than those of any other agricultural product. In 1997, some 500 government employees administered the regulations involving milk, according to the *Wall Street Journal.*

Federal milk price supports date to the 1930s, when dairy farmers lobbied for protection, arguing that in the case of a market reversal, they could not store their product until prices recovered, unlike, say, wheat producers. In response, the federal government created a formula to guarantee dairy farmers a minimum price that milk processors would pay.

In the 1960s, regions of the country remote from the Dairy Belt of Wisconsin and Minnesota were facing milk shortages. Congress reacted by passing the Federal Milk Marketing Order, widely known as Eau Claire Rule. It divided the U.S. into 31 regions and paid a bonus to farmers based on how far their cows were from what was then the geographic center of the dairy industry: Eau Claire, Wisconsin. Proponents argued that producers in the upper Midwest were more efficient and could produce milk at a lower cost, but because liquid milk was difficult to ship any great distance from that region, it had to be produced locally in other regions of the country. A price bonus would ensure the availability of fresh milk nationwide.

The system meant that Chippewa Valley producers re-

USDA, *quoted in the* Wall Street Journal, *November 25, 1997.*

The lowest price the Agriculture Department will allow processors in these cities to pay farmers for beverage-grade milk in December 1997:

City	Min. Price Per Gallon	City	Min. Price Per Gallon
Eau Claire	$1.19	New York	$1.37
Minneapolis	1.21	Charlotte	1.37
Chicago	1.22	Atlanta	1.37
Kansas City	1.27	Boston	1.38
Denver	1.34	Tallahassee	1.41
Dallas	1.37	Miami	1.46

ceived the lowest prices in the nation for their product. However, it was probably good for the country's consumers, since it enticed farmers located far from the Midwest to enter the dairy business.

As the years passed, the Eau Claire Rule came under increasingly serious attack from Wisconsin and Minnesota dairy farmers. In 1990, the Minnesota Milk Producers Association brought a lawsuit against the United States Department of Agriculture (USDA), contending that this method of computing milk prices was no longer valid. In 1997, U.S. District Judge David Doty of Minneapolis ruled that the existing geographic differentials could no longer be used to determine Class I milk prices. A little more than two years later, on January 1, 2000, the 31 regions of the U.S. regulated by a Federal Milk Marketing Order were reduced to 11 regions by the USDA.

In 1993, California surpassed Wisconsin as the nation's leading milk producer. Some people blamed the milk pricing system. But California didn't fall under the Eau Claire Rule. Federal Milk Marketing Orders regulate about 70 percent of U.S. milk. Those regions of the country that aren't subject to a Federal Order, like California, may have a state milk marketing order or they may be unregulated.

Working the Land

We cleared a lot of land; it was in timber. We cut timber and took out the stumps and made fields out of it.
  — Harold Kringle, interviewed 2000, speaking about the mid-1940s

Working the fields remained a central part of the farm operation even after wheat had given way to milk as the Chippewa Valley's chief agricultural product. Cattle eat hay, corn, and other feed crops, and many dairy farmers grow these on their own farms. Like any other aspect of farming, growing field crops involves knowledge, skill, and inventiveness.

In the first years after the Chippewa Valley was settled, a farm family planted, cultivated, and harvested what few crops they could by hand. Shortly after, the much greater yields made possible by clear land and power machinery naturally encouraged cooperation among neighbors. But by the 1950s, even greater mechanization allowed a family to once again handle its own fieldwork. When Jim and Joyce Solie bought their combine in 1958, they were the next-to-last among their neighbors to do so. The remaining farmer threshed one more year, but with no neighbors to help, the next year he gave in and bought a combine, too.

John Kysilko teaches family friend Marilyn Larson of Colfax how to operate a tractor on the Kysilko farm in the Town of Arthur, Chippewa County, 1942. Courtesy of Jeanne Andre.

[In 1977,] we started with 18 [cows on] 160 acres. So we had extra crops at that time. Now we have eighty cows and probably 300-some acres of crop land so we're actually short on feed now, where at that time, we supplied all our own feed and plus a little extra.
  — Don Moos, Chippewa County, interviewed 2000

Exchange of Labor

You went … and cultivated corn for somebody, they would come back when you were haying or something and, you know, it was more or less swap work. There were never dollars.
— *Don Foiles, interviewed 1998, speaking of the 1950s*

Before machinery allowed one farmer to do the work of many, an informal arrangement helped people accomplish larger tasks, and allowed them to have fun doing them. In the late nineteenth century, "exchange work" as farmers often called it, was often done within ethnic groups, but farmers came to find that their work was the same regardless of their backgrounds.

Until the middle of the twentieth century, Myron Wathke remembers, the men would get together for "thrashing, silo filling, corn shredding, sawing wood and that stuff. The women went a lot back and forth helping each other, when they had these crews coming, to make meals. There was an exchange of help between the women, too." In settlements throughout the Chippewa Valley, Amish farm families carried the tradition of exchange work into the twenty-first century.

The Amish have remained profitable even in such terrible times, by not running to the bank like everybody else is doing and getting bigger and bigger, by really maximizing their on-farm resources, and through community and cooperating.
— *Produce grower Steve Clark, in a keynote address to the St. Croix Valley Greens' Living Green Conference, 1992, quoted in* The Country Today

Charles McMahon's threshing crew, 1946, Pepin County.

Threshing the Neighborhood

If it was threshing time, you went over and helped your neighbor and then when it come your turn, he'd come by and help you. So that brought the whole community together.
     — Earl Myers, Menomonie, interviewed 1998, speaking of the late 1930s and the 1940s

In 1950, near the end of the threshing era, nine Menomonie area farmers went together to buy a threshing machine, paid for by a note from the Kraft State Bank. At the end of each threshing season, each farmer "paid-in" about a nickel a bushel for the grain threshed on his family's farm. In four years they'd paid off the note. They used their thresher together, and stored it in a shed they built together. In other neighborhoods, one person owned a thresher. Byron Berg's father brought his thresher to farms around Hale beginning in 1928. Farmers paid Berg about two and one-half cents per bushel and cooperated on the work to run it.

Before the 1880s, area farmers spent weeks threshing grain with hand tools, alone or in small groups. The threshing machine processed a whole field in a few days, but it took a crew of a dozen or more. "It was a long day," Myron Wathke remembered. "If you weren't on the road with your team by seven o'clock in the morning, you were late ... and it was all of that at night when you got home."

Theodore Kruger's threshing crew, 1913, east of Elk Mound, Chippewa County.

How a Thresher Worked

A thresher was a large machine used in harvesting small cereal grains. After a crop of wheat, oats, or barley was cut, a family would make bundles from a group of stalks and prop up a bunch of bundles into a cone-shaped "shock" to dry. After it had dried for some days, the grain would be "threshed." Machine threshing involved enormous labor, but it was a dramatic improvement over flailing grain by hand, which meant, simply put, beating it with specialized stick. A flail usually had a long handle and a shorter, free-swinging stick attached to its end.

To thresh, workers pitched grain bundles into the machine. A conveying chain carried those bundles into the separator. In the separator, the grain passed through a slot between a rotating cylinder and a grooved plate. Teeth on the cylinder beat the grain into the plate, knocking the grain's kernel free, but not crushing it.

The separated kernels, the chaff, and the straw then traveled over a rack, coarse enough to allow the grain and chaff to fall through. As it fell, a fan blew the lighter chaff and dust away from the grain. At the end of its descent, the loose grain hit an elevator, which transported it into a grain wagon parked nearby, or into individual bags, whichever the farmer wanted. Meanwhile, conveyors inside the thresher carried the straw out the rear of the threshing machine, where it was blown onto a straw stack.

The threshing machine remained largely unchanged until the 1920s when the the combination harvester appeared. Commonly called a "combine," it could cut and thresh at the same time. Although the combine slowly replaced the thresher, many threshers remained in use into the 1950s.

Typical threshing jobs

A farm family would cut and bind its grain, and stack the bundles into rows of shocks for drying. On threshing day, a crew of **pitchers** walked from shock to shock and tossed the grain bundles onto wagons circulating the field.

My older brother, Gunvald, every time another child was born, he would be very, very upset because he though that our family'd become so large and all the large families were so very, very, poor. . . . And when one was born, Gun was pitching [bundles] to my dad. . . . He was trying to knock my dad off by pitching so hard, because another kid was born and we'd become so poor. . . ..
— *Eric Fossum, Rock Falls, interviewed 1999*

The **bundle haulers** arranged the bundles on the wagons and drove them to the thresher, which sat stationary in the field. Two bundle wagons pulled up to the thresher at the same time, one on either side. The two haulers took turns tossing bundles into the thresher with a pitchfork.

> *Bound to be a lot of horse play . . . for instance, tying the last bundle down to the guide wagon so when he come to unload, he could fight trying to unload that last bundle.*
> — *Vernon Peterson, Siren, interviewed 1998, probably speaking about the 1930s*

The **machinist**, sometimes called the **thresherman**, maintained and monitored the thresher, and the steam engine or tractor that powered it. His post was at the engine or tractor, so it could be stopped in an emergency. **Machinist's assistants**, or **blower-tenders**, stayed with the thresher to record the amount of grain being threshed, and guided the distribution of grain into the wagon and straw onto the stack. The best could produce a tight, symmetrical straw stack from their perches, without a separate straw stacker.

> *I used to get to go with the threshing crew because my dad had the threshing machine.... We had an old truck with a grain box and so I had to stay in the truck and be sure all the grain was away from the spout that came down, so that it didn't back up and fill up with grain and plug up ... and boy, you were so dirty when you came in, because you were with that dust all the time.*
> — *Bernice Sutliff, Menomonie, interviewed 1998, speaking about the late 1940s or the 1950s*

The **straw stacker**, if there was one, shaped and compacted the straw as it was delivered out of the thresher. As the dirtiest and most demanding job on the crew, it paid a cash wage in some threshing circles, and would almost certainly be rotated among the members of the circle.

> *Stacking the straw was the dirty job, but the straw stacker received the best wine.*
> — *Clarence Werner, Weston, interviewed 1998, speaking about the 1950s*

The **grain haulers** drove the grain by wagon from the thresher to the granary. The machine filled the wagon or truck, then the haulers drove it to the granary and unloaded it with a scoop shovel. In other cases, the thresher was used to dump grain into individual sacks, not wagons: no scoop shovel needed.

The **catering crew** brought water to the threshing crew out in the field, and provided the workers with ample meals. The wives and daughters of the threshing circle, sometimes younger sons, made up this crew. The woman whose grain was being threshed provided the food and headed that day's crew.

In the horse-drawn wagon era, it was expected that the host farmer would provide grain and water for the **horses**. The horses often earned unthreshed grain bundles for their effort.

Feeding The Crew

You wanted to be sure that you didn't cook what the lady before had cooked, the day before, so you would ask your husband, "What did you have?" So you'd have a different kind of meat, different vegetables....
    — *Elaine Schroeder, Augusta, interviewed 2000, speaking about the 1940s*

Before she was married in 1956, Esther Bandli helped her mother prepare meals for the threshing crews that traveled from farm to farm. Esther usually served meals to "smaller" crews of eight to ten men, but did serve as many as twenty workers.

Children often helped with the lighter chores at threshing time, brought food and water to the fields, and learned the process by watching their friends and family. For women, threshing time helped maintain social relationships, as they often worked with family members or neighbors. It also allowed some women to showcase their culinary talents.

There'd be like, fifteen to twenty farmers there. And of course, all the wives would be making the meal for the day.... And I'd look at these guys while they were at the table and I knew they were very hungry because they'd take cobs of corn and run them like a typewriter across their mouth and the corn is just flying everywhere.
    — *Bruce Donaldson, Eau Claire, interviewed 2000, probably speaking about the 1950s*

Anna Kysilko cutting pies for a threshing dinner while her granddaughters Ann and Terry Andre look on, about 1955. Courtesy of Jeanne Kysilko Andre.

Pie pan, Stasko farm, Stanley area, Clark County, early 1940s.

Twilight of the thresher

Cadott farmer Don Foiles bought a threshing machine in 1952. He also bought a tractor, his first. He threshed eight or ten neighboring farms with his tractor and thresher. His tractor made the circuit without him at cultivating time. "I can remember the first tractor we bought went around the neighborhood from farm to farm cultivating. 'Cause it was the only tractor around."

Over the next ten years, he said, fewer and fewer farmers needed his machinery or his threshing services. "There was tremendous change from World War II on through the early seventies. There was money in agriculture and it brought a lot of people individualism; you know, they could get their own equipment and with good management, they could pay for a lot of that stuff that they couldn't have afforded earlier." As a result, he said, "everybody went their own way."

By 1960, combination harvesters, or "combines," were used almost exclusively to harvest small grains in the Chippewa Valley. The combine required a family-sized crew of two or three. Agricultural professionals actively promoted combines and criticized threshing rings as unnecessary and overly labor-intensive. But Carol Kringle of Barron remembered the threshing crew as a way to bring the neighborhood together and thought the combine brought one of the biggest changes she saw in the farming community. Modern machines also took serious money to buy and skill to maintain.

Combine maintenance break while harvesting a soybean field between Cornell and Cadott, Chippewa County, October 23, 2003. Left to right: Dave Sonnentag, Allen Sonnentag, Jamie Close. The Sonnentags, who ran a custom harvesting business, were hired to help with the harvest. Photographer: Scott Schultz. Courtesy of The Country Today.

What used to be a real simple solution, sometimes needs to be plugged into a computer now just to see that the battery cable is corroding or something.
— Tim Dotseth, Menomonie, interviewed in 2000

The Family in the Fields

Everybody pitched in. Mother would, you know, drive tractor. I don't know if she ever drove the horse. But we all helped when it was haying time and when they needed extra help, everybody pitched in and, and did something.

— Bernice Sutliff, Menomonie, interviewed 1998

[Mother] could drive the horses on the hay fork and could be trusted to do it well, so when they were putting the hay in the barn … they'd shout directions from the hay barn as to when she should stop, when the hay bales were in there, or the hay, and then she'd turn the horses around and go back.

— Phyllis Berg, Hale, interviewed 2000

Karo syrup pail later used as a lunch bucket, 1920s.

When he was young, Otis Fossum used to follow his dad around the field. "I'd walk behind him all day long when he was out on the plow or anything. When I was really, really small, I'd run ahead and put my feet in the [soil] and let the plow cover them up and watch the birds come and eat the worms." As he got older, he took over the plowing.

On a dairy farm, field work takes up much of the year. Farmers plant in the spring, cut hay and cultivate crops in the summer, and harvest corn and other grains in the fall. Haying and harvesting are especially busy times. Bruce Donaldson put up 2,000 bales of first-crop hay, 1,000 bales of second-crop hay, and 1,000 bales of straw after combining his grain. Such busy times have often required help, from relatives or a neighbor boy.

By 2000, the milking part of the operation at the Felix farm near Stanley had grown more complex. Eugene Felix made the choice to hire all of his field work out, even though the crops are still grown on his land.

When [our sons] were younger … we used to take them out when we baled hay. We'd take them out there with a bunch of toys, and they'd sit in the back of the station wagon [or with] the tailgate down in the pick-up… and they'd play with their toys, and [Ilene] and I'd be out baling hay.

— Don Moos, Barron County, speaking of the 1970s

Gathering oat bundles left by the grain binder to make into shocks, about 1921. Matt Schreir far left, Marie Schrier on grain binder, Schreier Brothers farm, near Stanley, Chippewa County. Courtesy of Millie Mickett and Stanley Area Historical Society.

Lunchbreak while shocking oats, 1920, Schreir brothers farm, near Stanley, Chippewa County. Courtesy of Millie Mickett and Stanley Area Historical Society.

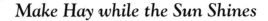

Hay carrier and crook, about 1927, used to carry loose timothy hay into the barn loft.

Make Hay while the Sun Shines

Timing is everything, as they say. Once hay is mown, it's left on the field for several days to dry before it's baled. Sunny days and clear nights are crucial. George Hayden of Pepin County would still use hay that had been rained on once after cutting, usually to feed the steers. If it was rained on twice, it was plowed back into the field — a total loss.

Farmers check hay frequently to see if it is ready to cut. Hay is at its most nutritious before it starts to seed out. If rain or high moisture delays the cutting, milk production will drop. Less milk equals a smaller milk check.

You ask any farmer if they'd rather have too much or not enough rain and they'll tell you too much every time. Because then there's something out there. They might have trouble getting it in, but there's something growing.

— Gary Evans, interviewed 2000

Loading hay into the hayloft, about 1937, Kubista farm, south central Washburn County. Courtesy of Rose Tomesh.

Horse Power to Horsepower

I could plow with the three horses, a 12-inch gang plow; the most I could get done was five acres a day…. You had to take an hour off for noon to give the horses a rest.

— Otis Fossum, Rock Falls, interviewed 1999, speaking of the 1940s

The introduction of new machinery eased the burden of individual tasks again and again over the 20th century. But stump-pullers, threshing machines, tractors, and combines also all let farmers accomplish more in a day. From 1930 to 2000 in Wisconsin, hay yield per acre doubled, milk yield per cow tripled, and corn yield per acre quadrupled. Harvesting those extra quantities demanded extra labor. So "labor saving" technologies didn't shorten the farm family's work day; they simply let the same work day produce more product. Buying and maintaining more machinery also meant spending more money.

It takes a lot of different equipment to make a farm go. Besides your tractors, you've got to have machinery to pull behind the tractors, all your tillage, planting, harvesting equipment…. I guess the haying equipment is the biggest equipment that we have the most of; it's your baler, your choppers, hayracks, and chopper boxes, and blowers.

— Larry Wathke, Fall Creek, interviewed 2000

Twilight of the draft horse

We bought one of them little F-12 International tractors in the winter of 1935…. That tractor made a farmer out of me. Because I don't give a darn, the horses would plain tire out, and oh, gosh that was terrible. You had all that work to do and then they'd stand there with their head down. You had to feel sorry for them, but that darn tractor — it wasn't very big, but it'd go night and day.
— Clarence Werner, Weston, interviewed 1998

In the 1930s Dorothy Bullis Carpenter's grandfather would be behind a team of horses and her father would be on a tractor — her mother would drive the tractor during haying season, or when the family was short of help.

The change from horse to tractor was gradual; it wasn't until 1954 that the U.S. Census of Agriculture ceased recording the number of horses per farm. But it was revolutionary. A person using a machine could do the work of several people, and several people with machines could do the work of a whole neighborhood.

The earliest tractors, steam-driven, helped operate threshing machines and other large equipment. Gasoline-powered, steel-wheeled tractors appeared in the Chippewa Valley in the 1920s. After World War II, rubber-tired tractors, and other improvements in machinery, allowed farmers to plant more acres, cultivate more effectively, and reduce the amount of time required to harvest an acre of crops.

In 2000, on the Weiss family's 300-cow operation near Durand, Don Weiss sometimes drove the semi truck hauling haylage (fermented hay stored in a silo) while his father drove the tractor. "And Don's … oldest boy is getting into it now," Don's sister-in-law Jan Weiss said. "He took tractor safety. He's going to be fourteen." Bruce Donaldson said that in his childhood, "the rule of thumb on the farm was, if you could reach all the pedals, it's time to learn how to drive it."

Louis Holfacker harvesting oats with a grain binder and four-horse team, about 1940, near Exile, Pierce County.

Son Bobby driving the family tractor for the first time, with father Vern Haas looking on, 1976.

George Hayden fixing equipment on his Pepin County farm, 2000. Photographer: Jeanne Nyre.

Use What You Have, Fix What You Can

If you are a mechanic you can take things apart, like my dad did, and some farmers I know are real good at keeping something going; they are keeping their machine costs down — and they can feed their animals.

— Dennis Stuttgen, DVM, Fall Creek, interviewed 2000

Leon Erickson tinkered the portable forge he used on his Eau Claire County farm so he could more easily use it to make and mend other equipment. He also modified a cream separator to run the forge.

Successful farmers become mechanics, welders, and tinkerers out of sheer necessity. Don Bensend bought his late-1940s-era welder from Lester Thompson who got it from the Barron High School agriculture department. Bensend used it to build wagons and repair equipment on his farm near Dallas.

Farmers also invent and improve existing tools to make work easier. Ronald Kohnke riveted a handle on a field-mower tooth to make a knife for cutting twine and corn stalks. Using leather straps and 20d nails, Paul Swoboda fashioned a "calf weaner," a spiked halter worn by a calf that keeps his mother from letting it suck. Machines shops also borrowed heavily from the house, recycling old cans and jars to store small parts and fasteners.

You fix what you can, and what you can't, you have to take downtown. But at $40, $50 an hour shop labor, and $10 [per hundred weight of milk] you fix what you can.

— Don Moos, Barron County, interviewed 2000

Handy Band Cutter

A BAND cutter is a time-saver around a corn shredder or where husked corn is being fed to the cattle.

The sketch shows a band cutter made by sawing a slot lengthways thru a piece of broom handle (about 7 inches in length) making the slot long enough to

Home-Made Band Cutter Made of Mower Tooth.

let an old mower or binder section slip in.

Next make holes in the handle to rivet thru and rivet the section in the slot.

Make a hole in the other end and insert a string and tie ends to hang it up by, when not in use, also to loop around wrist.

Twine cutter, about 1945, fashioned from an old sickle blade and handle from another tool.

Left: Instructions to make such a tool were printed in Handy Andy On the Farm, 1922, a collection of innovative ideas that had appeared in the monthly Farm Mechanics Magazine.

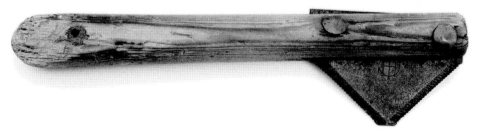

71

Risky Business

We had a team hitched up, and there was a colt of one of the horses on the team that ran between two of the three horses and got tangled up in the harnesses of the horses. And at that point, the team bolted and ran away and my father got dragged in the machinery behind the horses and had his leg broken very badly. That was during one of the bad drought years. So the rest of that summer, he laid in the house by a window, looking out at the dying fields, in the heat, unable to work, and just had the hired men to do the farm work. So that was a hard summer for all of us.
— *Margery Kohlhepp, Eau Claire County, interviewed 1998, speaking about the 1930s*

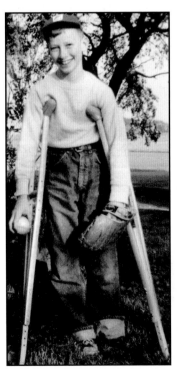

Terry Kohlhepp the day he was released from the hospital after recovering from a partial leg amputation as the result of grain auger accident, July 1962. Courtesy of Elmer and Margery Kohlhepp.

Like any wholesale business, farming is subject to financial and political risks: price volatility, economic downturns, shifts in government policies and the environmental and social concerns of the public. Like any manual labor job, it's subject to environmental hazards, such as chemicals and dust, and physical dangers, such as injury or overexertion. Like any skilled trade, it involves using wicked tools and machinery under great power and torque. Like any outside work, it's accomplished at the mercy of Mother Nature. Farming is like commercial fishing in that it combines all of these risks into one way of making a living.

I always had my chores to do, no matter if I had schoolwork or not, my chores always came first…. I remember back when I was in second grade I broke my collarbone. And my job was to sweep the feed alley up to the cows and string out some hay, and I told my dad I couldn't sweep because of my broken collarbone. "Well," he said, "use your other hand, then."
— *Jeff Pagenkopf, Elk Mound, interviewed 2000, speaking about his childhood in the 1960s*

Occupational Hazards

I had to go out and examine that pulley and, why, that's about that time the horses decided to go…
— *Vince Jesse, interviewed 1998, describing how he lost his finger during a hayloading operation, 1920s. He was five years old.*

At ten years old, Margery Kohlhepp's son Terry fell into a new grain auger at their family farm. He had to have his leg amputated. He had pushed a button to start the auger, which brought feed to the cows. As he was walking along the feed bunk, he fell in. The family wasn't aware what a dangerous invention it was.

Accidental death and dismemberment have been farming's most dramatic dangers. Nationally in 2003, crop farming had a worker death rate of 37 per 100,000 workers, compared with 23.5 for mining (though still far short of the rates for fishing and logging). But the biggest risk was respiratory, brought on by years of inhaling dust, feed residue, and chemicals. Farmers were three times more likely to develop respiratory diseases than the general population. Still, in 2004, 18 percent of the state's dairy farm families had no insurance at all, and less than 60 percent had health plans that covered all members of the household. Cost was a deciding factor. Said Ken Smith of UW–Madison, "If a farm family tries to purchase [health] insurance on the open market, the cost can be staggering — well over $10,000 a year."

Tornado damage near Colfax, June 1958.

I've seen farmers keep at it until they were too crippled up to do it any more....
— Richard Hughes, Eau Claire Leader-Telegram, 2003

Acts of God

Mary and I was married in '44 and we lived [on our first farm] until '58. Until the '58 tornado. It didn't leave a board where that 11-room house set. A neighbor boy died in the barn; we was both in the barn. He was helping me. He had just come to work for me that morning. The nicest kid in the county.
— Dan Emmerton, interviewed 1998, speaking about the 1958 Colfax tornado, one of only three F-5 twisters in Wisconsin in the last 50 years

While farm families are subject to the same economic stressors as any other group in the U.S. economy, they are also subject to hardships completely out of their control, hardships that urban dwellers often feel only in their lawns and gardens. Not only do natural disasters — tornados, wind sheers, hail — take their toll, but too much or too little rain, snow, heat, or sun can ruin a growing season or even bankrupt or destroy a farm.

Nineteen-eighty I remember very distinctly because I was picking up nails for the whole summer after that came. It was a straight wind that blew about 115 miles an hour and it just took a lot of stuff. And a couple other years where, where things just went drought. There wasn't enough moisture and you had to take what you could get and try to keep the cows going and we had enough but I remember the droughts especially. So when it rains like it did this spring, then it keeps coming, I don't complain about it raining too much.
— Bruce Donaldson, Eau Claire, interviewed 2000, speaking about the July 15, 1980, wind sheer and other weather problems in Eau Claire County

Learning to Farm

Agricultural Fairs are becoming an institution of the country; they are useful in more ways than one; they bring us together, and thereby make us better acquainted, and better friends than we otherwise would be…. But the chief use of agricultural fairs is to aid in improving the great calling of agriculture, in all its departments, and minute divisions — to make mutual exchange of agricultural discovery, information, and knowledge.
— *Abraham Lincoln, address before the Wisconsin State Agricultural Society, September 30, 1859*

In June 1957, the doors to Dunn County School of Agriculture and Domestic Economy closed for the last time. When "Aggie," as it was called, was founded in 1901 as the first agricultural high school in the United States, it joined the University of Wisconsin at Madison as a leader in agricultural studies.

In the nineteenth century, agricultural societies educated farmers through agricultural fairs. Just after the Civil War, Madison offered agricultural studies and began educating Wisconsin farmers through its "short course" in 1886. By 2002, the short course had about 5,000 living alumni.

Farmers can't apply herbicides or pesticides without a permit, so "you've got to go to school," said La Verne Ausman. George Hayden of Pepin County took a five-year farm training class, but he said he has also gotten a lot of information from magazines, including *Wisconsin Agriculturist*, *Dairy Today*, and *Hoard's Dairyman*, which he reads cover-to-cover.

Tim Bandli of Tony said he believes in learning from his peers. "You find out some about what the other guys have done," he said, "and that helps more than the information in the magazines. If you find out someone down the road tried this and it worked for them, then at least you can think about it."

Attendance is amazing, to me anyway…. My classes start at 8:30 at night and go until 11:00…. I'm working with dairy farmers mostly so I can't start before 8:30…. We run it according to when they need it run.

— *Gary Evans, Chippewa Valley Technical College,*
interviewed 2000, about his class on Farm Business and Production Management

FFA emblem, from Jason Wathke's high school FFA jacket, 1980s. The National FFA Organization (called Future Farmers of America until 1988) was founded in 1928 to support agricultural education. In 2004, FFA had 476,000 members in more than 7,700 chapters across the U.S. Wathke wore this emblem on his "blue jacket," a familiar symbol for FFA members. Although no one knows how many people have been members of the organization, more than 3 million blue jackets have been sold since 1933, when delegates to the national FFA convention adopted it as the official dress.

Preserving the Land

We have not yet gathered up the experience of mankind in the tilling of the earth.
— Dr. L. H. Bailey, Cornell University horticulturalist, 1911,
in his preface to Farmers of Forty Centuries *by Franklin Hiram King*

In 1953, Eau Claire County held the National Plowing Contest, which high-lighted scientific methods of tillage including contour farming and strip cropping. Inspired by the contest, a non-profit corporation began Wisconsin's Farm Progress Days the next summer, to be held in a different county each year. The first year's event included a pageant covering a century of tillage from oxen to tractors.

By the late 1860s, Wisconsin lawmakers already recognized that the depletion of forests by logging led to widespread flooding and erosion. In 1897 the legislature initiated a program to monitor and preserve the state's natural resources.

But farming led to an even more dramatic increase in erosion than logging had. Wisconsin soil erosion peaked in the 1930s, and farmers began adopting soil con-servation techniques. The Depression-era Works Progress Administration planted trees on Byron Berg's farm near Hale to slow wind erosion.

Farmers also enriched soil by fertilizing it with manure, minerals, and chemi-cals, and experimenting with how they rotated crops. More recently, agricultural experts introduced such conservation techniques as "no-till," which leaves last season's plant residue on the surface of the field — that is, it's not tilled under.

Our son, who does what we call limited till, he doesn't spend much time out there. He can handle 400 acres of corn in less hours than we handled 50.
— La Verne Ausman, Elk Mound, interviewed in 2000

The slightly lower yield from the no-till field [less than 2 percent compared to conventional yields] is offset by the economic benefits, including decreased work hours, less fuel and machinery used, and total amount of soil conserved. Over four tons of soil per acre were saved collectively on the farms.
— 2001 Report on the Multi-Agency Land and Water Education Grant Program, University of Wisconsin–Extension, on two demonstration farms in Chippewa County

Erosion, near Silver Mine Drive, Town of Union, Eau Claire County, mid-1930s. Cattle and horses used to crop the grass on the steep hillside, and the sandy soil couldn't hold. The fields have since grown into woods. Photo courtesy of George Johnson.

Silos and Science

Nineteen-twelve, put up the [stone] walls for the barn and silo on the Frank Barkley place. That was a big job. I would do my chores at home, then walk up after breakfast, start work at 7:00 a.m. and work until 6:00 p.m. I got thirty cents per hour. It took 50 days to do the job.

— William Meisegeier, Barron,
from his reminiscences

In the 1910s, the Wisconsin Dairy Association spread dairy farming information throughout the state by holding informational meetings. At the same time the College of Agriculture at the University of Wisconsin (Madison) began developing farmers' institutes to encourage farmers to use the college's scientific research on their own farms. The tower silo design was introduced by University of Wisconsin agricultural scientist Franklin Hiram King.

The silo itself is an ancient technology (the word comes from the Greek *siros*, a pit or a hole for storing corn), and ensilage is an ancient way of preserving animal feed, mentioned in the Old Testament Book of Isaiah. It seems to have lost favor from the first century A.D. to around the beginning of the nineteenth century. In contemporary use, silage means fermented, high-moisture forage, such as corn, hay, or oats. Silage is made from the entire plant, not just the grain. To ferment the plant material, it's harvested when it has a suitable moisture content, chopped, and then packed tightly into the silo to create an anerobic (airless) environment, which aids fermentation without letting the crop spoil. Silage maintains more plant nutrients than if the crop were dried. Silage is especially suitable for dairy cattle, which thrive on high-nutrient diets.

Scientific ideas including the silo faced initial reluctance from farmers. But by 1924, more than 100,000 silos dotted the Wisconsin landscape. Mechanization of farms also helped the silo gain ground — Mayland Heath remembers his father getting a labor-saving silo unloader on their farm near Tony "to keep his boys interested" in farming.

Top: Filling silo, Chippewa County, September 1939. Photograph by John Vachon. Library of Congress, Prints & Photographs Division, FSA-OWI Collection, fsa 8c36026.

Bottom: Filling silo at the John Winter farm, Eau Claire County, 1968.

Like other farming equipment, silos held hidden hazards. In January 2001, Darwin Mason was severely injured when he became caught 40 feet off the ground in a silo loader on his Chippewa County farm. A month later, Al Douglas of Elk Mound lay for seven hours trapped under a 2,000-pound chunk of frozen silage in his concrete silo. That July, still in a wheelchair, he completed a local mile race in just over twelve minutes.

The silo motor is quite a handy thing. I don't know of anybody around here that throws the silage out by hand anymore. Just for what they cost, they are so efficient and they take care of a very hard back-breaking job, especially in the winter when the silage is frozen.

— Eugene Felix, Stanley, interviewed 2000

Reshaping Lives

As more farm products are raised than can be used by the [Dunn County Asylum], the very considerable surplus is marketed, the receipts from the source providing an important item in its resources.

— History of Dunn County, *1925*

Walk through Orchard Cemetery on Eau Claire's west side and you'll see the unassuming concrete gravestones of hundreds of Chippewa Valley farmers slowly fading with time and rain. Among the corps symbolized by the flat markers, it's hard to tell which individuals farmed. Some stones have only names, not even birth or death dates. Some stones have not even names. But there are farmers among them; this was the cemetery for the Eau Claire County Asylum and County Home, and at one time most of the clients of those two institutions farmed.

Wisconsin inaugurated a system for keeping its "unfortunate" citizens in county asylums in 1881, to give them a "healthful occupation … tending to direct their minds into normal channels." Under staff supervision, patients raised crops and vegetables and maintained working dairy herds. Milk, cream, and butter were sold to local creameries. In 1945, the Eau Claire County Asylum and County Home showed a net profit of $25,780, more than twice that year's national average for farms of all kinds.

The Dunn County Asylum managed over 1,000 acres in 1925 and the Chippewa County Farm still maintained 700 acres when it closed in 1987. The Wisconsin Home for the Feebleminded, which sat across the river from the Chippewa County Farm, "was one of the first farms [of any kind] in northern Wisconsin to utilize modern dairy improvement practices."

Binding grain at the Eau Claire County Asylum, about 1900.

Other attempts to apply farming as a social instrument were not successful. In 1887, Congress passed the Dawes Act, which divided reservation lands into individual farms. At both Lac Court Oreilles and Lac du Flambeau, reservation agents hired "Farmers" who encouraged the Ojibwe men to clear their allotted lands and plant crops like potatoes, timothy, rye, and corn. Boarding schools at Lac du Flambeau and Hayward each maintained a farm. Their government "Farmers" instructed students, who did most of the work.

But reservation lands in Northern Wisconsin were not well-suited to agriculture, a fact often noted in agency reports. Ojibwe families survived by adapting, blending new ways and old. Some earned income from farming, although few had more than ten acres cleared. Gardens, maintained by women, were more common. Some continued to maintain beds of wild rice and process maple sap into sugar and syrup. But, to supplement their incomes, many Ojibwe sold their land to the lumber companies; on some reservations, over 90 percent of the land passed out of Ojibwe hands.

Inset, above: Ojibwe school children, Lac du Flambeau, about 1899. Agency buildings and the boarding school, which maintained a farm, can be seen behind them.

Left: Home of An du Kwe (Mrs. Carrie Alberts), Lac du Flambeau, May 13, 1922. Lac du Flambeau Industrial Survey, Roll 3, Neg. 269, National Archives, Great Lakes Region, Chicago. An du Kwe and her family owned one cow, three ponies, one horse, 32 chickens, and a two-acre garden. The survey praised her: "A widow. Keeps a bakery. Very industrious."

Gathering Together

Well, I don't even know that my great-grandparents ever learned how to speak English, because I think that they moved into a community that was their friends and relatives, and they could go to church, they could go to the stores, and communicate with the neighbors in the native language.
— *Eugene Felix, Stanley, interviewed 2000, speaking of the time between 1880 and 1920*

The Perovsek family came to Willard from Slovenia in 1908. With his neighbors, Frank Perovsek helped organize the Willard Cooperative Dairy and the North Hendren Cooperative Dairy, and helped build the buildings that housed them. He also had a hand in building the Holy Family Catholic Church, where he worshiped, and the North Willard school, where his children went.

In addition to the natural beauty and ancestral ties their farmsteads may offer, farm families are tied to the land by a sense of community. This attachment is born and maintained at peoples' homes and in farm work done together, but also, importantly, at gathering spots in rural towns. The general store may have become the convenience store, the saloon the sports bar, and the farrier the auto mechanic, but the stories are still about mighty white-tail bucks and the jokes about Ole and Lena.

Pleasant Valley Homemakers Club members showing off the party hats they had made in their monthly meeting, early 1950s, Chippewa County. Courtesy of Bev Peterson.

Investing in Each Other

. . . the middleman's the one who gets it all.

— from the song "The Farmer Is the Man," Farmers Alliance Songbook, 1890s

Top: Ridgeland Cenex feedmill co-op, 1950s. Courtesy of Wisconsin Farmers Union.

Bottom: Ridgeland Cenex, 2004. Photographer: Frank Smoot.

As immigrants settled the Chippewa Valley and began to farm, they had to pay high prices to have goods brought in. But they got little money in return for their produce, and had little control over its marketing, sale, or distribution. As early as the 1840s in Wisconsin, farm families formed cooperatives to protect their interests, and also to accomplish goals no family could accomplish alone, such as road-building, rural electrification, or insurance against disaster.

Cooperatives sprang from traditions brought over from Europe, and many were formed within ethnic settlements. However, based on the values of self-help, self-responsibility, democracy, equality, equity, and solidarity, cooperatives soon bound farmers together across ethnic lines against large "outside" forces, both corporate and governmental.

They also became civic organizations. The Washington Co-operative Creamery near Eau Claire — which began business in 1890 — had an auditorium with seating for 300. Beyond a space for its member meetings, the co-op opened the room for community Christmas parties and other local functions. In 1902, the co-op joined with eleven other country creameries to form the Eau Claire Creamery Company, which, in addition to its business, helped educate boys and girls about farming and sponsored annual farm events for FFA and 4-H members.

You take this little Ridgeland down here. If that didn't have that Cenex store, feedmill, and all, there wouldn't be no town.... That little town would have dried up. It's just a booming little town now. The only reason it's there is because of that co-op.

— Harris Kahl, Prairie Farm, interviewed 1998

Building a Cooperative

In 1942, the Colfax Co-op Creamery satisfied its mortgage and gained the right to burn the papers. The whole Village of Colfax turned out to celebrate with a parade and barbecue. The cooperative had been in operation since 1905, and built its new "modern" building in 1929. Members had to take out a mortgage since obtaining railroad rights had taken most of the previous 24 years' profits.

A cooperative is an organization started, owned, and run by the people who use its facilities and services. In the Chippewa Valley, many area businesses and utilities have been community-owned cooperative enterprises, including feed stores, creameries, cheese factories, feed mills, insurance agencies, and fueling stations; electric and telephone utilities; and egg-selling, ice-cutting, silo-filling, livestock trading, grain exchanging, and threshing services.

Although not all cooperatives succeeded, some have grown into political and economic powerhouses. The Wisconsin Energy Cooperative, a federation of eighteen central and western Wisconsin energy cooperatives, had a membership of 158,000 in 2004. Cenex Harvest States Cooperative, organized in 1931 as the Farmers Union Central Exchange, grew into a Fortune 500 company with 2003 sales of more than $9 billion.

What we did was form cooperatives, put our products together, and … marketed our products through cooperatives…. People who work for General Motors call themselves businessmen; employees is what they are. We are the businessmen, we are the keyholders, we are the capitalists.

— Mike Taft, Osseo, interviewed 2000

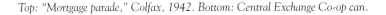

Top: "Mortgage parade," Colfax, 1942. Bottom: Central Exchange Co-op can.

"The Farmer Feeds Them All"

Many organizations have sought to organize farmers. Among the oldest of these were the Grange, formed in 1867, and the more-radical National Farmers Alliance, formed in 1877. In the 1890s, the Farmers Alliance took the lead in creating the Populist Party to "restore the government of the Republic to the hands of the 'plain people'." One of its rallying cries was "The farmer feeds them all!" While the party itself was short-lived, many of the its platforms have since been adopted into law.

Wisconsin farmers seem to have preferred more moderate organizations such as the Farm Bureau. State and county farm bureaus across the country, which date to the 1910s, helped fund county agricultural agents and promoted scientific agriculture. In 1919, these farm bureaus joined into the American Farm Bureau Federation, which came to be known simply as the Farm Bureau. Wealthier commercial farmers comprised much of the Farm Bureau's membership, and the federation had corporate ties. In other parts of the country, farmers distrusted the Farm Bureau and its business-class constituency.

By 1920, both the American Society of Equity and the Farmers' Union, which had grown out of cooperatives, moved into lobbying and collective bargaining. During the Great Depression of the 1930s, more radical organizations including the National Farmers Organization and the Farm Holiday Association organized farmers against corporate interests, and promoted political actions such as protests and "milk strikes."

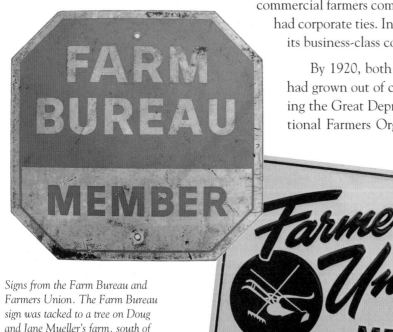

Signs from the Farm Bureau and Farmers Union. The Farm Bureau sign was tacked to a tree on Doug and Jane Mueller's farm, south of Eau Claire, Eau Claire County.

While the goals of these organizations sometimes seemed to be at odds with one another, many farmers belonged to more than one. As a young man in the 1930s, Harold Tomter of Pigeon Falls belonged to both an auxilliary of the Farm Bureau — to learn "how to make a living" — and the Farmers' Union, which was "attempting to get you a better living."

Float of the local chapter of Wisconsin Association of the American Society of Equity at the Inter-County fair, Stanley, 1908. Courtesy of Stanley Area Historical Society.

Neighbors: Help when You Need It

If it was threshing time, you went over and helped your neighbor and then when it come your turn, he'd come by and help you. So that brought the whole community together.

— Earl Myers, Town of Springbrook, Dunn County,
interviewed 1998, speaking about the 1930s

Sunday morning, May 17, 1953, tractors began to roll onto Bob Peterson's Chippewa County farm. Peterson, normally strapping at 6'3" and 240 pounds, had been laid low by rheumatic fever, and his friends and neighbors decided to help him with his spring field work. Plows, disks, harrows, and a planter roamed the fields. A crew cleaned out stock pens and another fixed fences. "I never knew I had so many friends," Peterson told the *Chippewa Herald-Telegram*. "It's a wonderful feeling to realize that others are so ready to help."

Neighbors plowing the field for Bob Peterson after he became bed-ridden with rheumatic fever, 1953. Six other tractors helped in the field that day, and additional crews cleaned the cow pens, fixed the pasture fences, mowed the lawn, and helped in the garden. Courtesy of Bev Peterson.

Neighbors helped each other complete large tasks, such as threshing or building barns. They even built civic structures such as churches and schools together. They also took on many of the civic duties we now think of as responsibilities of our county, village, or city governments — including fighting fires, grading roads, putting in poles for power lines, building bridges, and finding ways to finance it all.

As technology advanced, families could complete their farm work with less help, and they felt more isolated from their neighbors. Still, such events as benefit dances and the day-to-day operation of small-town or rural volunteer fire-and-rescue departments suggest that the spirit of mutual assistance survives.

When we first moved into Bloomer ... we were only there six months and a tornado pretty much took down most of our buildings and left just the house and the greenery. And boy, we learned our neighbors real quick.... It was just, just a devastating situation. Threw us all together and we realized all we had in common.

— Ilene Moos, Bloomer, interviewed 2000, speaking of a tornado in 1977

Neighbors working together to shovel a path wide enough to get the snow plow in to finish the job, 1936, northeast of Baldwin, St. Croix County. Courtesy of Myron Hesselink.

Chore coat worn by Bob Donaldson on his farm near Eau Claire. It was a gift from the Town of Union volunteer fire department.

Raffle ticket for the Rock Falls Volunteer Fire Department annual fundraiser, 2003.

Rural Rescue

In 2000, more than 50 fire fighters from four departments joined neighbors and family members to battle a barn blaze on Wayne and Cheryl Bowe's Town of Tilden farm. The barn, full of hay and straw in late August, was a total loss. But the firefighting effort saved the Bowe's four silos. The Bowes had twice before lost barns on the same site, by fire in 1930 and high winds in 1933.

After a disastrous 1881 fire in Durand, its citizens started a volunteer fire department. But other parts of the Chippewa Valley had to rely on neighbors with buckets well into the twentieth century. At some point a village or township board authorized an equipment purchase, and a fire department was born. Recent additions to area departments include such high-tech tools as SCBA gear and thermal imagers for "seeing" through smoke.

To finance such technology, departments are often awarded Federal Emergency Management Agency (FEMA) grants, but many departments also hold local fundraisers. In 2003, the Rock Creek Fire Department held a Pork Feed and Dance, at which they raffled cash and other prizes. In 2004, United Fire and Rescue, serving Hammond, Baldwin, and Woodville, planned a pancake breakfast, smelt fry, auction, and car show.

Although modern firefighting and rescue efforts are organized, rural residents still rely on their neighbors. The 1999 roster of the New Auburn Area Fire Department included a farmer, a butcher, a grader operator, a school teacher, a truck driver, a lawn-mower repairman, a writer, and a mother of five. These trained amateurs assembled, whenever the need arose, to save what they could of their neighbors' houses, barns, and fields.

Courage, a good neighbor and quality medical services made all the difference for Ed Hanson, 78, who was injured Oct. 15 [2003] in a horse-driving accident on his farm north of Baldwin. Mr. Hanson retired in 1993 but continued working on his farm. Without an active dairy to maintain, he's been able to work the land with horses instead of machinery.
— The Country Today, 2003. *Hanson's neighbor Randy Baumann, who was driving past, stopped when he saw the horse team pulling a plow without a driver.*

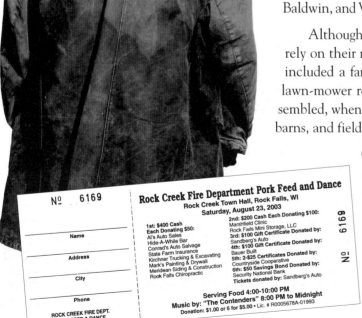

No 6169

Rock Creek Fire Department Pork Feed and Dance
Rock Creek Town Hall, Rock Falls, WI
Saturday, August 23, 2003

1st: $400 Cash
Each Donating $50:
Al's Auto Sales
Hide-A-While Bar
Conrad's Auto Salvage
State Farm Insurance
Kirchner Trucking & Excavating
Mark's Painting & Drywall
Meridean Siding & Construction
Rock Falls Chiropractic

2nd: $200 Cash Each Donating $100:
Marshfield Clinic
Rock Falls Mini Storage, LLC
3rd: $100 Gift Certificate Donated by:
Sandberg's Auto
4th: $100 Gift Certificate Donated by:
Bauer Built
5th: 2-$25 Certificates Donated by:
Countryside Cooperative
6th: $50 Savings Bond Donated by:
Security National Bank
Tickets donated by: Sandberg's Auto

Serving Food 4:00-10:00 PM
Music by: "The Contenders" 8:00 PM to Midnight
Donation: $1.00 or 6 for $5.00 • Lic. # R0005678A-01993

Name
Address
City
Phone

ROCK CREEK FIRE DEPT.
PORK FEED & DANCE

No 6169

Fighting the barn fire at Wayne and Cheryl Bowe's farm, Town of Tilden, 2000.

Meeting Places

[In our neighborhood,] we had the Sampsons and the Kearns and the Agasetts, and Torgersons and Rasmussens, Halversons, Jacobsons. Pretty much all Norwegians.... I remember we, we had a picnic one summer, 19 — oh, about 1906, I think. We had about, oh, must have been 65 people there, kids and all, at the Petersons ... and Jacobsons — they all got together — Olsons.... They all came with horse and buggies, and some of them walked.

— *Melvin Christopher, Menomonie, interviewed in 1998*

The hamlet of Allen was so small that, unlike most Chippewa Valley communities, it didn't have a church or a tavern. "But then," local resident Art Nix told the *Eau Claire Leader-Telegram*, "if it didn't have a tavern, it didn't need a church." People got together at the John Deere farm implement dealership. "Those John Deere Days were pretty good," said Nix, "with free beer and free food. There would be cars and trucks all over town for that. It was one of the biggest events of the year."

Dairy Days parade, Barstow Street, Eau Claire, 1956 or 1957. Photographer: Davis Studio

Taverns, halls, cafes, churches, cooperatives, volunteer fire departments, cemeteries, and team sports leagues were generally supported by the same people in a rural community. Those people hosted, staffed, and attended fairs and festivals. Events such as the Eleva-Strum Broiler Festival and Augusta's Bean and Bacon Days brought farmers and non-farmers together and were often used to introduce farmers to new ideas, methods, and equipment.

As time passed, better roads and cars allowed people to shop in cities, technology allowed a family to accomplish work it once took a neighborhood to do, and "outsiders" began moving in to rural neighborhoods, using them as bedroom communities or refuges from the urban world. Despite these changes, rural gathering places still formed an economic system and a human network that sustained rural life.

Eau Claire County 4-H fair held in Augusta, 1929.
Courtesy of Augusta School District.

Margaret Russell with her calf at the Eau Claire County
4-H fair, Augusta fairgrounds, 1925.

Bright Lights

In the first years the Chippewa Valley was settled, families rarely travelled off the home place. When they did, it was usually to church, less often to town. Once or maybe twice a year, the town would shine under the torches or lights of the county fair or a passing circus. For young people of the Valley's early days, this might be their first experience with the larger world beyond the boundaries of the home place or the neighborhood.

Sometimes combined with carnivals, pageants, and other activities, fairs have long had a complex relationship with their audiences. On the one hand they were family fun and the place for neighbors to engage in fellowship and friendly competition, from four-wheeled-wagon backing demonstrations to pie-making — and pie-eating — contests. On the other hand, they could expose visitors to unsavory elements. They were good wholesome fun in daylight, an illicit adventure at night.

The rabbit barn at the Rusk County 4-H fair, Ladysmith, 2003. Photographer: Kathleen Roy.

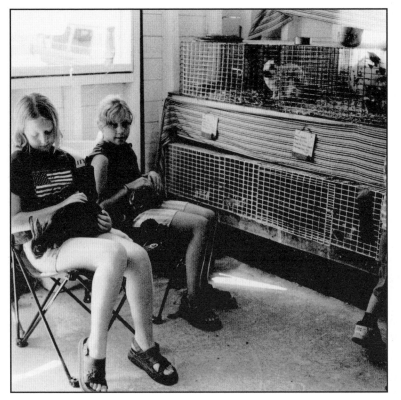

Besides bringing a little piece of the city to the country, however, the fair also brought the country to the city — an opportunity for city dwellers to see farm animals, tractors, and other trappings and activities of rural life.

First organized by agricultural societies, agricultural fairs were promoted by extension services, land-grant colleges, and local trade-center merchants to display, demonstrate, or even introduce ideas and products relating to farming and rural life. Fairs let farmers show off their wares and teach the non-farming public about their economy and the goods and services they offered. Fairs also gave a chance for young people to improve and demonstrate their skills in a wide variety of endeavors, including art, mechanics, science, riding, and animal husbandry.

I got my chores done and I didn't change overalls either, and that, she never forgot that. And she come in the restaurant and I was sittin' there and I says, "You know what I think? I think we should go to the Menomonie Fair tonight." She says, "I think we should, too." And that's how we got started.

— Harris Kahl, Prairie Farm, interviewed 1998, on courting his wife of 56 years

A selection of rosettes and ribbons from the Eau Claire County Junior Fair, the Eau Claire County Holstein Breeders, Northwest Wisconsin District Fair in Chippewa Falls, and the Rusk County Junior Fair. "Premium" winners, such as represented by the ribbon at left, took home cash as well as honor.

Russells Corners 4-H Club, Augusta fairgrounds, Eau Claire County, 1922.

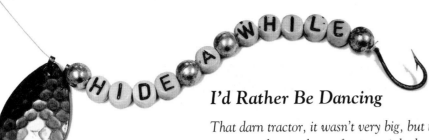

I'd Rather Be Dancing

That darn tractor, it wasn't very big, but it'd go night and day as long as I wasn't going to dances. If I went to a dance, then, of course, it had to sit still.

— Clarence Werner, Weston, interviewed 1998,
remembering 1935 when his family got their first tractor, a Fordson

Above: Fish lure made for the Hide-A-While bar, 1990s.

Below: Annual Christmas celebration at the Hide-A-While bar, south of Rock Falls on the Dunn and Pepin County line. The tavern opened in 1963 in a one-room school that had closed the previous year.

During World War II, when he was just out of high school, Don Foiles met his wife at the Rainbow Gardens near Cadott. The dance hall was separated from the liquor bar, and all ages were welcome at the hall. "They had county inspectors that stood right at the door, and I don't know of anybody ever being turned away because they were too young.... Lots of people would come, families with their small children would come and they were always welcome at those places."

Country taverns and dance halls took hold in the Chippewa Valley in the ninteenth century. Often featuring local brews and homemade food from nearby farms, these rural institutions also sponsored sports teams, employed area residents as cooks and musicians, and provided venues for wedding dances and other important celebrations. Today, fewer taverns and dance halls are part of the rural landscape, but in 2004 some still thrived as centers where celebrations and traditions help tie the bonds of the rural community.

More so than in some other parts of the country, area ethnic traditions have included alcohol, although not all farm families approved of its use, and there has long been a temperance movement in Wisconsin. Eau Claire writer Waldemar Ager published *Reform*, a Norwegian-language temperance newspaper, from 1896 until he died in 1941.

We used to have Christmas Eve at the house here. Everybody would come on Christmas Eve. But ... the kids kept getting bigger and bigger and more kids, you know, grandkids, and so now we have it over at Club Ten, and we rent the back room.

—Jan Morrow, Cornell, 2000

Interior photographs of Sokup's Tavern, 1950s on Hwy. 53 north of Rice Lake, Barron County. Courtesy of Ralph Sokup.

Teams, Clubs and Associations

My mother was a city girl. But she adapted very well to the farm. She was real involved in everything, and she was a 4-H leader, and she was a woman that could get things done.
— Bernice Sutliff, Menomonie, interviewed 1998, probably speaking about the 1940s

In 1950, Dunn County had 25 Homemaker's Clubs. During the year's program of organized events, more than 150 women learned about making braided rugs, testing water hardness, serving foods rich in vitamins A and C, and understanding soil types. They were charged with passing the information along to non–club members.

Neighbors organized clubs, groups, teams, and associations that were social, recreational, and often educational. In 2000, Eugene Felix of Stanley belonged to the local FFA Alumni and the Friends of the Library Group. "So, we do have certain things that we say are for the community good," said Felix, "but we do it partly to socialize with people, too."

In the first decades of the twentieth century, the Zapadni Ceska Bratrska Jednota (ZCBJ) held classes to cultivate the "mother tongue" and to teach English to newer immigrants at the Bohemian Hall north of Cadott. Fraternal members also organized gymnastics for young members. Through the middle part of the century, the Hall held wedding ceremonies, receptions, anniversary celebrations, funerals, birthday and card parties, polka dances, and country, bluegrass, and gospel concerts. The hall fell into disrepair, but a 1999 drive brought new Hall members, and non-member volunteers lent helping hands to restore and preserve the Hall. Since 2000, the building has hosted a "Long Live the Squeeze Box" program, a *jaternice* (sausage) and potato pancake breakfast, and *kolache* classes (*kolache* is a sweet, yeasty bun), among many other events.

Baseball game on Jim Roycraft's farm, late 1920s, near Cadott, Chippewa County. Courtesy of Lois Roycraft Krumenauer.

The boys, I got them started in horseshoe when they were young. The stakes are still out here yet and we are all playing horseshoe yet in a league…. Pretty tough bunch to beat.
— Myron Wathke, Fall Creek, interviewed 2000

Mock wedding of Steve and Rose Najbrt in honor of their thirtieth wedding anniversary, 1934. The event was held at the ZCBJ–Western Bohemian Fraternal Association, Lodge #141, locally known as the Bohemian Hall, near Cadott, Chippewa County. Mock weddings were a common Bohemian tradition. Courtesy of Jeanne Kysilko Andre.

Harry Anderson's farm, Town of Wheaton, Chippewa County,
1930s. Big Elk Creek Lutheran Church stands in the background.
Harry was the son of Lars and Grethe Anderson, who settled the land
in 1857. Courtesy of Irene Ovren.

Joining Hands

Our church, we consider, is our family.

— Lucille Schumacher, Elk Mound, interviewed 2000

Irene Ovren and her husband Roald moved to the Chippewa Valley after World War II. They lived in a log home on a farmstead in Chippewa County, within view of the Big Elk Creek Lutheran Church. Roald Ovren was buried in its church yard in 2001, as have been members of each generation living in that house, which was built in the 1860s. The home held the first meetings of the congregation, formed by Norwegian Lutherans.

When Harris Kahl was a child near Prairie Farm in the 1920s, "Going to church was just like eating breakfast," he said. "There was no 'ifs' and 'ands' about it; we all went to church." For Chippewa Valley farm families, the churches have been the places they celebrated community and maintained their sense of identity. Churches were often the first community buildings that neighbors built, to serve as places both of worship and of fellowship, holding community celebrations and rites of passage.

Dedication celebration of the Methodist church in Vance Creek, Barron County, 1909. Courtesy of Barbara Bender.

Building the Church

[Our] house was built in 1917, and then the church … was built in 1920 by all the members of the congregation.

— Bernice Sutliff, Menomonie, interviewed 1998

Recently, the cornerstone of Prairie Farm's United Methodist Church was unearthed and opened. Inside rested a list of those who'd donated to the building of the church. In this town of 600, some surnames of the contributors were still familiar in 2004: Rassbach, Kahl, Miller.

In 1906, the Beef Slough Lumber Company tore down its recreational center and gave the lumber to the Ladies Aid Society for the purpose of

building the Union Free Church in Nelson. When the Free Church burned in 1942, the Society purchased the Spring Creek Church near Durand, and had it renovated and moved to the Free Church site. Members of the congregation dug a basement for the structure and put on a new roof.

At first, rural congregations in the Chippewa Valley divided themselves by ethnic group as well as by denomination. A Norwegian Lutheran church might sit near a German Lutheran church. But more recently, as membership has declined, congregations have consolidated. Many have now even disappeared. St Raymond's, near Brackett, joined together three former parishes.

Although ethnicity could divide communities, churches, as community centers, could also bring them together. In 1999, a New Auburn teenager was killed in an auto accident. Although she was Lutheran, the much larger Catholic church in New Auburn opened its doors for the Lutheran funeral, so all the mourners could be accommodated.

It was difficult for the people… they were baptized there, educated there, married there, and their children were baptized there, and they certainly wanted to be buried there and that wasn't possible.
— *Lenore Krajewski, interviewed 2002, on the 1991 closing of St. Anne's parish near Bruce*

Church Events

Lucille Schumacher's church, St. Joseph's Catholic Church in Elk Mound, hosts an annual Fall Festival. Each year, individuals are responsible for a different job — clean-up, serving, or working in the kitchen. The festival features a dinner, bake sale, auction, polka mass, and cakewalk. For the 2000 Festival, Lucille contributed six pies, eighteen coffee cakes, and ten loaves of nut breads to the sale. The money raised through these efforts supported the church.

Our Savior's Lutheran Church in Rice Lake invented its popular Ole and Lena Suppers to bring together the congregation and the general public. Food traditions still associated with a particular background of the congregation may be among the last remaining expressions of ethnic identity in many communities.

The parish depends on the income from these events to keep it alive, and it's just good socializing when it comes down to it.
— *Peggy Kahler, Rock Elm Methodist Church, Elmwood, Eau Claire Leader-Telegram, 1997*

Baptismal font, Our Redeemer's Lutheran Church, 1886. The rural Eau Claire County church was closed and dismantled in 1976.

Reaching Out

Since 1956, the Ladies Aid Society at St. John's Lutheran Church in Spencer has been providing Christmas gifts to servicemen and shut-ins. In November 2000, members of the Chetek Lutheran Church loaded 658 quilts onto a truck to be distributed through Lutheran World Relief, which operates in fifty countries. They had sewn the quilts over the previous year along with 128 baby layettes — and school, health, and sewing kits.

Through 2004, Amish women gathered at each other's homes around Augusta as part of a quilting bee group. The money earned from selling the quilts has been put into a community pot to help pay hospital bills for members of the community.

In 2003, the *Superior Catholic Herald* noted that members of St. Anthony Abbot Parish in Cumberland were part of an interdenominational group dedicated to helping migrant workers. Seneca Foods, operating a green bean processing plant in Cumberland, employed the workers, many of whom were from Mexico. Starting in 1998, St. Anthony Parish provided a location where people could bring in donated food, clothing, household goods and furniture for the migrant workers. The group also began helping find housing for workers newly arrived. Until 2002, St. Anthony held summertime Spanish masses complete with a Spanish choir.

Doesn't that make you feel good?

— *Donna Fults, Chetek,*
Eau Claire Leader-Telegram, 2000

Top: Diners at Our Saviour's Lutheran Church's annual Ole and Lena Lutefisk and Meatball Dinner, Rice Lake area, Barron County, November 2000. Courtesy of George Theis.

Bottom: Quilters from Our Savior's Lutheran Church, Rice Lake area, Barron County, 2000. Courtesy of George Theis.

Pete Schumacher walking his daughter Betty to the bus stop on her first day of kindergarten, 1972. Courtesy of Lucille Schumacher.

Taking the Bus

I would have loved to have gone on to [high] school. Nope, didn't have school buses then. I'm the oldest of ten in our family, and the first five, none of us got to high school. The last four all got to high school. They had school buses then.
— *Priscilla Laufenberg, near Alma Center, interviewed 1998, speaking of a change that took place around 1930.*

When Phyllis Berg of Hale went to high school around 1940, at first one of the boys drove out from town to fetch her and her neighbors. But, she says, "Then they decided that the school bus would be great. So Adler Thompson had a truck.... He made a school bus with seats in it for everyone. It was a pretty interesting vehicle. So, we rode the school bus. He became rather famous for his vehicle."

Before 1960, rural Wisconsin children lived within walking distance of their elementary school, although for some it was a long walk. In 1955, nine-year-old Christopher Owen had to walk more than two miles to Sunnyview School in the Town of Washington, as did his young siblings.

School consolidation took children away from the farm, literally and figuratively. Consolidation offered all Wisconsin children, urban and rural, an equal educational standard. But the disappearance of the rural schoolhouse hurt rural neighborhoods by lessening the interaction of their neighbors. Military service, high school and college educations: These were paths off the farm for many young adults, but it was not always a permanent departure. After his military service, Harold Graff seriously considered not coming back to his parent's farm, but he found that the career he wanted, electrical engineering, was saturated in the late 1940s. He returned home, and five decades later, passed the farm to his son Doug.

Jeanne Kysilko and neighboring farm kids ride the bus to Cadott, 1940s. Courtesy of Jeanne Kysilko Andre.

The One-Room Country School

Before we went to school, we all spoke Slovenian. We spoke Slovenian at home. As soon as our older brother and sister started school, they started English, and we learned it from them. We were talking English pretty good before we started school. But the first ones — the oldest ones — they didn't know English when they started school. It was hard for the teacher, because the teacher didn't speak Slovenian.

— Josephine Trunkel, Willard, interviewed 2001, speaking of the late 1920s

End of the school year clean-up, Big Drywood Creek School, Chippewa County, 1920s. Courtesy of Jeanne Kysilko Andre.

One Monday morning in the winter of 1934-35, in Eau Claire County's Sunnyview School, teacher Daisy Mason reached in her desk drawer for a match and found her papers chewed into a nest for four baby mice. In December 1937, a skunk trapped under the school discharged its scent. "The smell had died down pretty well by the time of our Christmas program," Mason noted. This was life in the country school.

The basic structure and activities of a day at Sunnyview did not change significantly from the 1880s well into the 1940s. The teacher taught all eight grades and conducted twenty class sessions during the day. Children could listen in on classes given to the more advanced students, and "learn ahead," or listen again to lessons they'd already had, and "catch up." Each student was given lessons appropriate to his or her abilities, regardless of age or grade level.

The goals of a country-school education had not changed much since colonial days: to transfer cultural knowledge and to teach life skills such as good health, moral conduct, and citizenship.

Halloween, probably 1942, Sunnyview School, Town of Washington, Eau Claire County

A Community Center

I know that the school was used when people for one reason or another needed a gathering place. As a matter of fact before my time I can remember my grandmother telling how the Methodists used to meet there and conduct their church services in the school house.

— Russell Doane, Menomonie, interviewed 1999, probably speaking of a time around 1920

Sunnyview School in Eau Claire County was a neighborhood project. In 1936, Mr. Lee brought a new flag and Mr. Smith fixed the merry-go-round. The next year, the Smiths built a woven wire fence around the school yard. Parents and neighbors provided wood for the stove and shoveled the school out after snowstorms.

Country schools were often near the center of public life in a rural neighborhood. Picnics, town meetings, 4-H clubs, and even wedding receptions would be held at schools.

And in many ways, schools were the most localized seat for all levels of government. In May of 1942, ninety-one people registered at Sunnyview for War Ration Coupons. The county nurse came regularly to give vaccinations and test the children's eyesight. On snowy days, the letter carrier left mail at the school for neighbors to pick up.

Left: Sunnyview School, Eau Claire County, after a snowstorm, January 1937.

Right: Rolland Fried square dancing for the school's annual Christmas program, Sunnyview School, Eau Claire County. Courtesy of Rolland Freid.

Roasting marshmallows after the end of the year clean up, Nobel School, Chippewa County. Courtesy of Verna Klemish.

School Consolidation

The day started out like any other day. The last afternoon they had a picnic of hot dogs and potato chips. No visitors. They played games outside until it started to rain. They ate picnic lunch inside. It was a very gray day.

— *Florence Teeple Beels, 1960,
on the closing of Hamlin School, Trempealeau County*

In the 1920s and 1930s, 6,500 one-room country schools dotted the Wisconsin landscape. Eau Claire County had 85 such schools in 1927. That number declined steadily as people moved into the cities and as "progressive" ideas took hold in education, so that in 1960 there were 1,300 one-room public schools in Wisconsin. But by 1970, according to the Wisconsin *Blue Book*, there were none.

By Wisconsin statute, all elementary school districts had to affiliate with high school districts by July 1, 1962. Certain schools would have to be expanded, and some students would have long bus rides. The aim was to provide an equal education for all Wisconsin children, no matter where they lived.

The remains of Fernwood School, Chippewa County, 2002. Photographer: Mary Gladitsch. Courtesy of Mary Gladitsch.

Some people criticised one-room schools for the uneven and sometimes inadequate preparation of the teachers, and the old-fashioned curriculum. Others defended the spirit of cooperation and the values that the small schools gave their students. Such issue could still arouse passions decades after consolidation. In 2003, rural Cleghorn School closed amid great controversy, including an effort to recall members of the Eau Claire Area School District board.

The Wisconsin Evangelical Lutheran Synod still operated thirty-six affiliated one-room schools in the 1983-84 school year. In various parts of the country, the Amish have built schools or even bought abandoned one-room schools at auction and returned the buildings to their original purpose.

Cleghorn School serves as a center for our community, and not having a school here would damage the fragile stability of an already weakened rural economy. We value our children and our rural way of life.
— *Jane Mueller, Fall Creek,
in a letter to the "Voice of the People,"
Eau Claire Leader-Telegram, 2002*

The Future of Farming

There probably will no longer be a "typical" Wisconsin dairy farm.
— Ben Brancel, Secretary of the Wisconsin Department
of Agriculture, Trade, and Consumer Protection, 2000

Seventeen-year-old Margaret Mueller, milking with a pipeline system, 2000. Photographer: Jeanne
Nyre. A pipeline system takes milk directly from the cow to a bulk tank.

Something Smells

City people ... especially if they haven't had a farm background, they don't understand these smells.
— Gary Evans, farm business and production management instructor,
Chippewa Valley Technical College, interviewed in 2000

It's a familiar moment to any summer traveler around the Midwest: Driving through the smells of newly plowed soil or fresh-mown hay, suddenly an invisible, acrid cloud of "farm" blows in through the car's windows or its air system. Experienced area residents can even tell turkey farm from hog farm from dairy farm by smell.

Tony Tomesh spreading manure on his farm, County M, northeast of Rice Lake, January 1991. Photographer: James Leary. Courtesy of James Leary.

The smell has become a symptom of larger issues, including runoff pollution, urban sprawl and land use. According to the Associated Press, by 2001 people building homes in rural Wisconsin areas had raised enough of a stink about bigger farms that produce more manure to prompt state officials to wonder what to do about the mess.

"It's a tough balancing act," said Wisconsin Agriculture Secretary Jim Harsdorf. "If Wisconsin doesn't resolve that issue, it could in fact harm the future of our livestock and cropping industries."

During the construction of a subdivision that called for $500,000 homes to be built adjacent to the south end of their property, Warren and Isabel Brown of Hartland, in Pierce County, painted a sign and nailed it to a hayrack: "We still spread manure."

I still like it in the winter time where I can walk into a nice warm barn and kick on the barn lights and get the smell of the cows eating hay.
— Steve Siverling, Bloomer, interviewed 2000

Runoff

There's not a farmer in the state who comes close to the concentration of fertilizer used by a home-owner [following manufacturer's instructions for using lawn fertilizer].
— *John Malchine, chairman, Board of Agriculture, Trade, and Consumer Protection, 2001*

In March 2002, an operator from John De Farms near Baldwin spread three million gallons of manure on 195 acres of cropland. But only the top three inches of soil had thawed and the manure couldn't be plowed into the field. Much of it ran off into Girard's Creek, a tributary of the scenic and trout-filled Kinnickinnic River. In 2003, the operation reached a settlement agreement with the Wisconsin Department of Justice, which included penalties of almost $43,000 for violating the manure management requirements of its Wisconsin Pollutant Discharge Elimination System permit.

After the passage of the federal Clean Water Act in 1972, Wisconsin made great strides in controlling pollution from industrial and sewage plants. By 2000, runoff from parking lots, golf courses, barnyards, crop fields, and residential lawns was the main factor in making one-third of the state's rivers and lakes unsafe for swimming and fishing. This kind of pollution is called "non-point pollution," since it comes from broad areas like fields and housing developments, rather than specific points, such as the outlet to a sewage treatment facility.

Runoff from animal wastes and fertilizer, and sediment buildup from erosion, pollute not only local streams and rivers, but, since area rivers flow into the Mississippi, they have contributed to an enormous oxygen-depleted "dead zone" in the Gulf of Mexico.

Cleaning the hog pen on the Bullis farm, south of Eau Claire, about 1920. Courtesy of Dorothy Bullis Carpenter.

Starting in October 2002, administrative rules required farmers to meet new standards for applying fertilizer, controlling soil erosion from cropland, and managing manure. Among other things, the rules required farms maintain grass buffers at least 20 feet wide along waterways bordered by crop land, barn yards or feedlots, if state grants were available to cover 70 percent of the cost of creating the buffers. The rules also covered runoff from urban developments, charging business owners, landowners and cities themselves with several tasks.

In most lakes around the state, putting in a 30-foot [grass] buffer strip can reduce the pollution by 90 percent. The grass acts sort of like a coffee filter.
— *Dane County Supervisor Brett Hulsey, 2002*

Land Use

We have more neighbors now as the city is moving to the country and new homes have gone up, and you don't see these people. I mean, you don't hardly know who they are any more—who lives where.

Carol Moen, Rice Lake, interviewed 1998

Don and Ilene Moos moved to Chippewa County from Illinois in the 1970s. "It was to the point that you took your life in your hands when you took a tractor on the road because the urban sprawl got to be, be so much," he said of his home state. "And the land values were so far out of sight we couldn't afford to continue down there."

Thirty years later, some farmers feared that the Chippewa Valley might be facing the same trend. If land lost to farming is developed into subdivisions, it brings more traffic, higher land values, which has often meant higher taxes, and neighbors who can be unsympathetic to the realities — sounds, smells, dust, chemical use — of farm life. Ironically, Doug Graff, owner of a farm just south of Eau Claire, said in 2000 that his new suburban neighbors were glad that his farm was still operating, since it served as a buffer between their homes and the further encroachment of suburban dwellers.

In 2002, Dennis Caneff, Midwest regional director of the American Farmland Trust, said Wisconsin's land was being converted to urban uses five to six times faster than population growth. Still, in 1997, about 45 percent of Eau Claire County's land, about 55 percent of Chippewa County's, and more than 65 percent of Dunn County's, was in farms or grassland. In all three counties, much of the rest was in forest.

A 1999 Wisconsin poll of farm families found that "In most cases, farmers are torn between their desire to protect agriculture from outside pressures and their ability to retain their rights to sell their farmland at appreciably high development prices if they decide to quit farming." The temptation to sell was tempered in 2000 when the Wisconsin Department of Revenue adopted a use-value tax assessment code, taxing lands used for agriculture at a much lower rate than nearby residential or commercial parcels. Supporters of the code argued that it would help keep farming economically viable in the state.

There are two ways of looking at it — you like to preserve the way things were all the time, but you realize that progress happens.

— Doug Graff, Eau Claire County,
Eau Claire Leader-Telegram, 2000

Amish buggies south of Augusta, 2002. Courtesy of Eau Claire Leader-Telegram.

Housing development bordering a pasture in the Village of Hallie, Chippewa County, February 9, 2002. Photographer: Shane Opatz. Courtesy of Eau Claire Leader-Telegram.

Stacked milk cans, a common sight in Wisconsin in the earlier decades of the 1900s, stand in contrast to a large cream tanker truck at Grassland Dairy Products, Inc., near Greenwood in Clark County. The milk cans were owned by an Amish farmer who lives near the creamery. Courtesy of The Country Today. *Photographer: Scott Schultz.*

Go Big, Stay Small, Get Out

Fewer and fewer Wisconsin farms look like the ones on rest-stop postcards. Scientific breakthroughs, governmental regulations, and market challenges have influenced the nature of farming and rural life. Some farmers elect to GO BIG with heavy reliance on technology. A less widely publicized alternative is *go small.* "Farming the government" (collecting payments for keeping land out of production), "agricultural tourism" (creating experiences for tourists), and "raising whitetail" (charging urban deer hunters fees to hunt) are but a few of the ways individual families have maintained a rural life. Behind these phenomena lies the larger contemporary debate over land use.

Growing beyond Tradition

In 1960, Walter Pagenkopf milked fifty cows in a traditional barn. In 2000, his son Jeff and daughter-in-law Marie milked 400 in a milking parlor with the help of nine full-time and three part-time employees. In 2004, the old barn still stood at the heart of the farm, surrounded by an expanding web of modern buildings.

So far, large dairy operations have remained a small part of the Wisconsin farming landscape. Of the 23,000 Wisconsin dairies in 1998, only 90 had over 500 cows. Most operations had between 30 and 100. But large dairy farms are here and here to stay.

The technologically advanced dairy farms have little in common with those of our imagination. In 2002, the Five Star Dairy near Elk Mound had 750 cows, twenty employees, and generated almost $2 million in gross revenues. Covered walkways spanned the buildings and a 9-million-gallon waste-storage pond occupied more than two acres. But the Five Star Dairy did not signal a corporate invasion; it began as a partnership of three local family farms.

A thousand acres — it's just a dream I've had ever since I was sixteen years old, is to own a thousand acres…. It's not that I want a big corporate farm…. It's, whenever you do anything, if you set a goal, a mark … it's something to work for.

— Dan Weiss, Durand, interviewed 2000

Leaving the Dairy Life

[My son Doug] said 'Someone bought the whole herd this morning. They're going to be gone by eight o'clock tonight.' And they were. You make the decision … get it over with.

— Harold Graff, Eau Claire, interviewed 2000

In 2000, Dan and Jan Weiss were planning to expand their dairy farm, as was the neighbor to the east of their place near Durand. But his neighbors to the west and north were quitting for health reasons, and the neighbors to the south wanted to "pursue a different angle in life," as they put it. Dan said, "There's a lot of neighbors that are saying, 'one more year, one more year'."

In 1955, 130,000 dairy farms made up much of the Wisconsin landscape. During 2000, the number dropped below 20,000. The state lost almost 40 percent of its dairy farms during the 1990s alone. While low wholesale prices and the consolidation of processors and grocery chains were often blamed, complex forces drove the decline.

Federal supports propped up milk prices and didn't give smaller family operations much incentive to update and increase efficiency; then supports fell away, leaving little safety net. The expense of expanding, which could be in the millions of dollars, was often beyond what small operators could afford. And Wisconsin's long tradition of small family farms biased both farm families and their neighbors against large dairy enterprises. Western dairy states, such as California, did not have this tradition.

Age, illness, natural disaster, or simple frustration could all sink a dairy farm. Increased land prices, especially in areas "close to town" and ripe for development, tempted some to sell the homesteads. But others who gave up dairying didn't necessarily give up the land. Some rented out the fields; raised field crops, beef cattle, horses, or exotic animals such as llamas, red deer, or bison; or simply retired and let the land go back to nature.

I was 62 when I retired. I sold the cattle and rented out the land. But now it's hard even to rent the land, there's so much of it. My son still farms the fields, buys the seed and hires out the labor. He's a retired police chief so he can afford to do it even if he loses a little money. A farm gets run down if you don't farm it. The fields dry out, and choke up with brush. And goldenrod. That's a bad weed. A pretty flower, but a bad weed.

Josephine Trunkel, Willard, interviewed 2001

Making It Work

At the turn of the millenium, area farmers faced two main choices about how to stay in business: go big and increase revenues or stay small and keep costs minimal. Kevin Mahalko began a rotational grazing program on his farm near Gilman in 1996. In rotational grazing, farmers subdivide pastures into small plots. Cows graze one area closely, while the remainder of the pasture rests and recharges. For Mahalko, the program not only lowered his production costs, but reinvigorated his love for farming. "You're right out there walking in the grass, handling the cows." Area Amish families have also succeeded by "staying small." In 2004, one in four Eau Claire County farms was an Amish farm.

Packaging from specialty dairy products produced in the Chippewa Valley: CC's Jersey Creme yogurt made at Cave Creek Jerseys farm in Spring Valley, St. Croix County; butter made at Midvalleyvu farm, near Arkansaw, Pepin County; cheese made from organic milk produced on Sweetland Farms near Menomonie, Dunn County.

Several farm families found yet another way to survive. They added value to the operation at the size it was: Some produced organic milk, which demanded a higher price at market; others made their own cheese, yogurt, candy or ice cream.

Leroy Clark and Michelle Wieghart of Cave Creek Jerseys near Spring Valley seriously considered expanding their herd in 1999, but instead concentrated on selective breeding, pasturing their cattle, following a strict nutrition and vaccination program, and maximizing cow comfort, so that each of their cows produced more and lived longer. In 2002, they built a plant in Durand to manufacture a line of Jersey Crème yogurts. "It's our solution to making a better living in the dairy industry," said Clark, "doing it without trying to get great big."

One Morning in the Chippewa Valley

September 2, 2004, a Thursday, was set to be very warm. Already 67° at 5 a.m., it was heading toward 85° or better. Not atypical, actually, in Wisconsin's Chippewa Valley. It was a day like hundreds of early September days over the past century.

That morning, Harold Graff woke up about 7:30, checked the weather both on the radio and television "because so many things you do on the farm is determined by the weather," and took a half-mile walk, an activity he'd just returned to after being ill earlier in the summer. Harold was retired and his son Doug had taken over the farm just south of Eau Claire.

Neighbors stopped by to get the chopper box and tractor from Doug, who would have been helping except that his accountant was arriving shortly. Then Doug's eldest son pulled in and Harold talked with him. Then Frank Carr came out to work a garden plot he maintained on the Graff farm. Then the accountant pulled in.

Harold Graff, Eau Claire County

This was the house and farmstead where Harold was born, where he's spent all his life save three years in the military. He went into partnership with his father in 1949, married in 1951, and lived on the land with his parents until they passed away. His mother and late wife Phyllis did most of the milking, while Harold and his father worked the fields. As his father had, Harold passed the operation onto his son, although until 2003, Harold still hauled hay from the fields to the barn. After that, "because of age and health," he restricted himself to hauling out the recycling, or taking the lawn chair to watch "anything major" such as a building project — "not to influence, just to watch." His legacy, he hoped, would be "that my farm practices preserved the soil to the best of my ability, so that the soil is there for future generations, if they so desire, to use for agriculture."

That morning, Judy Gilles set herself about the business of running Cabin Ridge Rides, a 400-acre recreational facility offering all kinds of fun but anchored by horse-drawn vehicle rides. The operation near Cadott has been refashioned from her ancestral farm. Her husband Rusty's great-great-grandparents were the first to arrive in the nearby Irish Settlement and started to log Paint Creek in 1853. Judy's great-great-grandparents arrived in the Irish Settlement by the early 1860s.

Judy Gilles, near Cadott

While she waited for her two-year-old granddaughter to arrive (Judy provided grandma daycare service), she went to the catering kitchen to ready cleaning supplies for a trip through the woods to the land's wedding chapel, where she would clean chairs in preparation for an upcoming ceremony.

She'd just gotten together with two friends from high school, now living in Arizona and Illinois, and was planning a trip to the Waverly Midwest Horse Sale in Iowa, where they would be camping with friends from all over the Midwest.

Her future, she said, was running Cabin Ridge Rides "as long as my health holds." Her friends with "regular" jobs were retiring. "Since my time is my own, except when I have a group on the premises," she joked, "I may already be retired." She considered her legacy to be the place where her granddaughter had been spending her days, where her daughter was raised, where she, her parents, and her grandparents were raised. "I hope," she said, "I can pass on a love and commitment to the land."

Lucille Schumacher, Elk Mound

That morning, Lucille Schumacher, "a very happy homemaker and farmer's wife," made breakfast for herself and her husband Pete, made sure he took his medicine, then picked up an 86-year-old neighbor to attend mass at "St. Joe's Church" in Elk Mound.

A granddaughter had been visiting recently from out of state. Over the past several days, they'd all visited with many relatives, including the girl's other grandparents in Rhinelander. As Community Outreach Chair of the Dunn County HCE (Home & Community Education), still affectionately known as "the Homemakers," Lucille was making plans to represent HCE at the state convention in Green Bay.

It wasn't an unusual day or week, although, as she said, "every day is different. Besides loving to bake and cook, I do bookwork. I volunteer many places. I take time to golf, visit relatives, friends, even go on some trips. I love to sit in my swing and watch the activities around the farm. I never wake up fearing a boring day."

Lucille is 70, her husband 79. They farm "not because we have to — our children are all raised and doing well — so now we farm for the fun of it and because we still love it." She considered her children as her legacy, and hoped "to leave our farm in better shape than we bought it, especially the land."

That morning, Margaret Mueller, Meg as she's known, hiked the 20 minutes from her apartment to the University of Wisconsin–River Falls. A senior finishing a Dairy Science major, Meg was off to her first day of fall classes. Afterward she stopped by the basement of the Chalmer Davee Library to pick up that semester's rental texts.

She'd been quite used to her summer schedule, working as a veterinary assistant at the Animal Care Center in nearby Hudson; and, only the day before, she'd been at the Minnesota State Fair with her roommates. So the classroom was quite a change of pace; "after a few weeks, however, I'll be back in the swing of things." In the next few days, she would be finishing her application to veterinary school.

Meg planned to be a large-animal vet, "deal with farming in that way." But, she said, "actually living on an operating farm is not out of the question either." In any case, she wanted to "still have ties with our family farm in some way," speaking of the dairy farm of her parents, Doug and Jane Mueller, between Fall Creek and Eau Claire. Doug's grandfather bought the farm in 1903. Meg's goal as a future veterinarian was "to educate people" and her legacy to perhaps "discover a new bacteria, or perfect a technique that will carry on my name!"

That morning, at Sandy Acres Dairy near Elk Mound, which he runs with his wife Marie, Jeff Pagenkopf was feeding cows.

Margaret "Meg" Mueller, River Falls

Jeff Pagenkopf, Elk Mound

Pages 113-115: Five photographs taken the morning of September 2, 2004. Photographers: Carrie Ronnander, Kathleen Roy, Frank Smoot.

Selected Books for Further Reading

Allen, Terese. *Wisconsin Food Festivals: Good Food, Good Folks and Good Fun at Community Celebrations*. Amherst, WI: Amherst Press, 1995.

Apps, Jerry. *Barns of Wisconsin*. Black Earth, WI: Trails Books & Prairie Oak Press. 1977, 1995.

_____. *One-Room Country Schools: History and Recollections*. Amherst, WI: Amherst Press. 1996.

Browne, W.P., Skees, J.R., Swanson, L.E., Thompson, P.B., & Unnevehr, L.J. *Sacred Cows and Hot Potatoes: Agrarian Myths in Agricultural Policy*. Boulder: Westview Press, 1992.

Danbom, David. *Born in the Country: A History of Rural America*. Baltimore: The Johns Hopkins University Press, 1995.

Gulliford, Andrew. *America's Country Schools*. Washington, D.C.: Preservation Press, 1984, 1991.

Gough, Robert. *Farming the Cutover: A Social History of Northern Wisconsin, 1900-1940*. Lawrence, KS: University Press of Kansas, 1997.

Leary, James P. *Wisconsin Folklore*. Madison: University of Wisconsin Press, 1999.

Neth, Mary. *Preserving the Family Farm: Women, Community, and the Foundations of Agribusiness in the Midwest, 1900-1940*. Baltimore: The Johns Hopkins University Press, 1995.

Pederson, Jane. *Between Memory and Reality: Family and Community in Rural Wisconsin, 1870-1970*. Madison: University of Wisconsin Press, 1992.

Perry, Michael. *Population 485: Meeting Your Neighbors One Siren at a Time*. New York: HarperCollins, 2002.

Pfaff, Tim. *Paths of the People: The Ojibwe in the Chippewa Valley*. Eau Claire, WI: Chippewa Valley Museum Press, 1993.

_____. *Settlement and Survival: Building Towns in the Chippewa Valley, 1850-1925*. Eau Claire, WI: Chippewa Valley Museum Press, 1994.

Vennum, Jr., Thomas. *Wild Rice and the Ojibway People*. St. Paul: Minnesota Historical Society Press, 1988.

Selected Area Histories

Bandli, Esther. *Pioneers Rest: A History of Halvorsen Homestead Community*, 2000.

Bullis, Jack G. *Before Oakwood: The Bullis Edgewood Stock Farm.* Eau Claire, WI: Heins Publications, 2000.

The Chippewa County Chronicle. Friendship, WI: New Past Press, 1995.

The Country Yesterday. Eau Claire, WI: Eau Claire Press Company, 1988.

Gladitsch, Mary Rufledt. *Remember When: A Tribute to the Vanishing Rural Landscape.* Woodruff, WI: The Guest Cottage, Inc., 2001.

Miller, Gene. *Fairchild… When You and I Were Young*, n.p., 1989.

Raihle, Paul. *A Valley Called Chippewa.* Cornell, WI: The Chippewa Valley Courier, 1940.

Spominska Zgodovina: Historical Memories, Willard, Wisconsin. Willard, WI: Slovenska Druzba, 1982.

Selected Primary Sources

Archival and Oral History collections

CVM Fields and Dreams Oral History Project, 1998-2000. Museum staff and volunteers completed a total of 62 interviews with 84 individuals during the course of this project, which was funded by the Wisconsin Humanities Council. Interview subjects included farmers, both retired and those currently on the farm, agricultural professionals, an agricultural lender, an agricultural science instructor, and people who remain on the "home farm" through strategies other than dairy farming. Ten Chippewa Valley Agricultural Extension county agents were interviewed about current farming practices and trends in their counties, with a focus on dairy, feed crops, and specialty crops.

CVM Sunnyview School Oral History Collection, 2002. This oral history project was part of the *Building A Life: Creating New Contexts for Historic Structures* project that focused on the research and re-interpretation of CVM's historic structures. CVM staff and volunteers interviewed thirteen former students of Sunnyview School, which served a neighborhood a few miles south of Eau Claire from 1882 to 1961. Students were interviewed about their first day at school, daily routine, recess activities, holiday celebrations, discipline, and rural life. The *Building A Life* project was funded in part by a grant from the Wisconsin Humanities Council.

CVM Rural Life Documentation Initiative, 2000. Funded by the National Endowment for the Arts, Wisconsin Arts Board, and Wisconsin Humanities Council. This project aimed to collect, document, and interpret Chippewa Valley folk arts that reflect rural and farm life. Folklorists conducted work in their area of expertise.

Gilmore, Janet. "Cook, Cook, Cook, Cook, Cook." Aesthetics of table, home and garden, needle arts, traditions for preserving and preparing food, celebrations.

Leary, James. "Getting Together at the Crossroads Cafe (and everywhere else)." Social gatherings for music, dancing, food, and games.

Olson, Ruth. "Calves in the Old Barn." Evolving architecture, new uses for old barns, the barn as a center for farm life and family interaction.

Glenn Curtis Smoot Library and Archives. CVM's library holds about 25,740 images and other archival materials. Materials available for *Farm Life* research and exhibit development include diaries from local farm families, a large photographic collection of images of rural and farm life, and documents, including maps, tax deeds, and certificates. CVM collected photographs and documents from farm families during the planning phase, and several long and short-term loans are held in the library.

Wisconsin Historical Society, Wisconsin Agriculturists Oral History Project. Five Chippewa Valley region interviews and related materials; interviews conducted by Dale Treleven of the Historical Society staff in 1975.

Wisconsin Extension Homemakers Council:

The Impact of Her Spirit: An Oral History. Wisconsin Extension Homemakers Council, Inc. 1989.

Taste of Wisconsin History: Oral History Cookbook Reader. Wisconsin Extension Homemakers Council, Inc. 1987.

Area Research Centers, Local Historical Societies, and Business Archives

Area Research Center, Northern Great Lakes Center. Sawyer County tax assessment rolls and personal property records, 1926-1945; Charter of the Farmers' Mutual Assistance Insurance Company, 1901-1937; reports of midwives and physicians, with numerical reports of birth and death dates, 1888-1903; interview with Tom McClaine, regarding operations of the American Immigration Company, 1906-1940; papers of Roy R. Meier, a Price County farmer and historian, 1881-1978; agricultural and dairy statistics, 1880-1906.

Area Research Center, University of Wisconsin–Eau Claire. A wide variety of primary sources reside at the UWEC Area Research Center, a branch archive of the State Historical Society of Wisconsin. Related collections include records from the Union Mortgage Loan Company, located in Eau Claire, Wisconsin. The company made loans to farmers in Wisconsin and records document agricultural credit, farming conditions, and extensive personal data on the individuals to whom it lent money and the physical conditions of their property. A collection of diaries of early farmers in the Chippewa Valley region, including Henry Barber, Pepin County, Orpha Ranney, Dunn County, and Lucy Hastings, Eau Claire County. Frank Meinen's papers, including financial and production records of his dairy farm near Chippewa Falls, Wisconsin from the 1920s through the 1970s. Papers also include records of his participation in agricultural organizations, including the Wisconsin Farm Bureau Federation and the Cooperative Union of America. Records pertaining to the operation of the Flambeau Valley Farms Cooperative, Ladysmith, Wisconsin, 1925 to 1980. The Cooperative collection includes corporate records, milk and plant production records, memos, promotional materials, loan application requests, and correspondence.

Area Reseach Center, University of Wisconsin–River Falls. Collections reviewed include an oral history collection compiled by the St. Croix County Extension Homemakers from 1973 to 1974. Topics addressed include farm and daily work, recreation, education, and family life.

Area Research Center, University of Wisconsin–Stout. Archives include the Norval Ellefson Collection. Ellefson, a dairy farmer from Dunn County, served on the board of the Cenex cooperative for 23 years. The collection includes interviews, a Cenex history, various newspaper clippings, and photographs. Records from the Dunn County Fair, 1894-1910.

Barron County Historical Society. The Barron County Historical Society houses a large collection of photographs relating to the agricultural development of the region. A review of this collection brought forth a number of photographs relating to the exhibit's themes of neighbor to neighbor, making do, and risk and recovery. Images include a wood cutting bee for an injured neighbor, the interior and exterior evolution over time of the Barron Cooperative Creamery, and documentation of a variety of specialty crops in the region.

Cadott Area Historical Society. The collection includes several oral history interviews that were reviewed for this project, as well as many images and documents relating to farm and rural life.

Chippewa County Historical Society. The society manages a large archive and library, with many items relating to agriculture and rural life, including collections on cooperative businesses, one-room schools, and individual farm families.

Chippewa Valley Electric Cooperative. The collection includes institutional and membership records from its founding to the present and many images documenting the growth of CVEC and the use of electricity in its service area.

Eau Claire Energy Cooperative. The Eau Claire Energy Cooperative has photographs and a complete collection of *Wisconsin REA News/Wisconsin Energy Cooperative News.*

Jump River Electric Cooperative. The co-op holds documents, unpublished manuscripts, and photographs relating to its organizational history and membership use of electricity.

Stanley Area Historical Society. The society provided many resources for the project, as it had recently completed an exhibit about agriculture and farm families in the area. Collections used included photographs, documents, and newspaper clippings.

Steenbock Agricultural Library, University of Wisconsin–Madison. A review of this collection found interesting documents and photographs relating to the history of the UW Extension Service and Extension Homemakers for several Chippewa Valley counties.

Wisconsin Farmers Union Archives. This archives includes scrapbooks, photographs, and documentary materials on WFU and other cooperatives dating from the 1930s to the present.

Wisconsin Historical Society. The iconography collection includes state extension education photographs and images of family farms, creameries and cooperatives, and specialty crops production.

Selected Web References

The Core Historical Literature of Agriculture
http://chla.library.cornell.edu

Eau Claire Press Company
Story Archive
http://www.cvol.net/ecpc/archive.htm
Photo Gallery
http://photogallery.cvol.net/default.asp

Electronic Texts for the Study of American Culture
http://xroads.virginia.edu/~HYPER/hypertex.html

United States Department of Agriculture
National Agricultural Statistics Service
http://www.nass.usda.gov
Wisconsin Agricultural Statistics Service
http://www.nass.usda.gov/wi/index.htm

University of Wisconsin–Madison
Center for Dairy Profitability
http://cdp.wisc.edu
College of Agricultural and Life Sciences
http://www.cals.wisc.edu

University of Wisconsin–Extension
http://www.uwex.edu

Wisconsin Historical Society
http://www.wisconsinhistory.org

Acknowledgements

Many individuals and organizations participated in the *Farm Life* project, which included a 5,000 square-foot gallery exhibition at the Chippewa Valley Museum, traveling versions of the exhibit and related publications and services. The Chippewa Valley Museum thanks all who gave their time, energy, financial support, creativity and enthusiasm. At the time of publication, the project was still in progress. This list reflects contributions as of September 30, 2004.

Funders

major grants
National Endowment for the Humanities, Division of Public Programs
> Planning, $41,400
> Implementation, $225,000
> Outreach, $112,000
Wisconsin Humanities Council
> Fields and Dreams Oral History Project, $9,646
> Building a Life: Sunnyview School and Anderson Log House Re-
> Interpretation
> Farm Life 2005 Museums Conference/Teachers Institute, $10,000
Wisconsin Arts Board Folk Arts Program
> Rural Life Documentation Initiative 2000-2002, $14,957
> Farm Life Exhibit Design and Construction, 2003-04, $9,025

gifts or grants of $10,000-$24,999
Janet Barland*
McDonough Manufacturing and the Tietz Family*
National Endowment for the Arts, Folk and Traditional Arts Program
Rural Life Documentation Initiative $10,000

gifts or grants of $5,000-$9,999
Barb and Marlow Wathke*
The Dairyland System of Cooperatives
Wipfli Ullrich Bertelson, LLP*

gifts or grants of $2,500-$4,999
Marv and Mag Lansing*

gifts or grants of $1,000-$2,499
Ralph and Peggy Hudson*
Dr. and Mrs. Robert M. Lotz*
RCU*
Tony and Nancy Schuster*
Kathryn Teeters*

gifts or grants of $100-$999
Accelerated Genetics, Adopt-A-Class Fund
Eau Claire Electric Co-operative, Adopt-A-Class Fund
Linda and Steve Erickson*
Harold Graff*

*Gift to the 25th Anniversary Capital Campaign designated as match for a grant from the National Endowment for the Humanities.

Chippewa Valley Museum Staff

planning and implementation
Karen DeMars, educator; Janet Dykema, director of community programs; Dondi Hayden, facilities manager; Susan McLeod, director; Jeanne Nyre, designer; Carrie Ronnander, curator; Kathie Roy, assistant curator; Frank Smoot, director of publications; Susan Sveda-Uncapher, assistant designer; Eldbjorg Tobin, librarian.

implementation
Dennis Erpenbach, maintenance supervisor.

planning
Sara Anderson, director of public programs; Susan Glenz, education coordinator; Leah Hamann, educator; Julie Johnson, curator of collections; John Pellowski, curator of public programs; Tim Pfaff, curator of public programs; Diane Schmidt, senior curator.

other
Dorie Boetcher, operations manager; Judy St. Arnault, office manager; Annette Truitt, volunteer coordinator; Wes Berg, operations assistant.

Consultants

academic

North Dakota State University: David Danbom, Ph.D., History.

University of Wisconsin-Madison: Janet Gilmore, Ph.D., Landscape Architecture and Folklore Program; James Leary, Ph.D., Folklore; Mark Louden, Ph.D., Linguistics (Amish and Mennonite History and Culture); Ruth Olson, Ph.D., Folklore, Center for the Study of Upper Midwestern Cultures.

University of Wisconsin-Eau Claire: Robert Gough, Ph.D., History; John Hildebrand, English; James Oberly, Ph.D., History; Jane Marie Pederson, Ph.D., History.

Independent Scholars: Cathy Inderberg, Mary Ellen Stolder, Fields and Dreams Oral History Project Coordinators.

University of Minnesota: Lary May, Ph.D, American Studies.

University of Missouri-Columbia: Mary Neth, Ph.D., History.

University of Missouri-Truman State: Steve D. Reschly, Ph.D., Amish and Mennonite History

Purdue University: Paul Thompson, Ph.D., Ethics and Philosophy

farmers

La Verne Ausman, Dunn County; Judy Gilles, Cabin Ridge Rides, Chippewa County; Doug and Harold Graff, Eau Claire County; Carol and Harold Kringle, Barron County; Doug and Jane, Margaret, Steven and Peter Mueller, Eau Claire County; Jeff and Marie Pagenkopf, Chippewa County.

specialists

Marcia Wolter Britton, Director, Wyoming Committee for the Humanities, museum education; Mary Korenic, Director of Educational Programming, Milwaukee Public Museum, museum education and visitor studies; Paul Martin, Head of Exhibits, Science Museum of Minnesota, planning; Michael Perry, writer.

multimedia

Bruce Burnside, musician and composer, musical direction; Douglas Smith, technical production; Virginia Smith, University of Nebraska-Lincoln, script consultation and direction; David Evans and Craig Theisen, Science Museum of Minnesota, multimedia production; Steve Terwilliger, Art Department, University of Wisconsin–Eau Claire, multimedia production; Larry Glenn, Media Development Center, University of Wisconsin–Eau Claire, multimedia production.

on-line exhibit production workbook

Ann Koski, Director, State Historical Museum, Wisconsin Historical Society; Lawrence Sommer, Director, State Historical Society of Nebraska; Chris Schuelke, Director, Otter Tail County Historical Society, Fergus Falls, Minnesota; Robert Teske, Ph.D., Milwaukee County Historical Society.

educators

James Jeffries, North High School; Becky Mattson, Robbins Elementary School; Judy Reinhart, Longfellow Elementary School.

Program Officers

Clay Lewis, National Endowment for the Humanities
Dena Wortzel, Wisconsin Humanities Council
Rose Morgan, National Endowment for the Arts
Rick March, Wisconsin Arts Board

CVM and CVM Foundation Boards of Directors, 1998-2004

Janice Ayres; John Bachmeier; Jill Barland; Kathy Bartl; Marty Fisher-Blakeley; John Bowman; Linda S. Clark; Sue Diel; Duane Dingmann; Robert Downs; Linda Erickson; Barb Fey; Chuck Forster; Evalyn Frasch; George Gagnon; Jacob Gapko; Jeff Halloin; Jan Haywood; Helen Heinz; Laurie Hittman; Sally Kaiser; Mary Kay Kopf; Pam Kozuch; Ann Julsrud; Robert M. Lotz; Yong Kay Moua; Dean Olson; Tom Pearson; Jim Pinter; Dan Riebe; Susan Rowe; Rick St. Germaine; Sue Sausker; Barb Schmitt; Mary Schoenknecht; Tony Schuster; Kaye Senn; Eric Swanson; Dean Schultz; Kathryn Teeters; Al Templin; Sonya Tourville; Ken Vance; Ronald Warloski; Barb Wathke; Becky Welke; Mike Welsh; Wayne Wille; Norb Wurtzel.

Oral History Projects

Building a Life: Sunnyview School

Interview by Janet Dykema: Betty Pendergast Bowe.
Interviews by Amy Benson: Charles Burce; Orville Lee; Jane Pendergast.
Interviews by John Wright: John Burce; Rolland Freid.
Interviews by Jayne Dooley: Ron Cater Arlene Hagen; Shirley Nicolet; Martha Zank Hummer.
Interviews by Paul Nagel: Ardis Crowe; Evelyn Krenz.
Interview by Carol Wick: Robert W. Smith.

County Agricultural Agents

Interviews by Janet Dykema and Meg Marshall: Bob Cropp, Pepin County; Carl Duley, Buffalo County; Jim Faust, Dunn County; Matt Jorgenson, Clark County; Jerome Clark and Randy Knapp, Chippewa County; Mark Kopesky, Price County; Mervin Markquart (farmer), Chippewa County; John Markus, Ashland and Bayfield Counties; Bryan Pierce, Vilas County; Kevin Schoessow, Sawyer and Washburn Counties.

Fields and Dreams

Interviews by La Verne Ausman: Russell Doane; Ethel and Tom Heath; Margaret Kent; Alyce and Earl Myers; Sharon Schaefer; John Weinzirl.
Interviews by Amy Benson: Otis Fossum; Eric Fossum; Cleve Kirkham; Margery Kohlhepp; Lenore Krajewski; Larry Wathke; Myron Wathke.
Interview by Jennifer Bauer: Byron and Phyllis Berg.
Interviews by Kari Bostrom: Melvin Christopher; Don Foiles; Jim and Joyce Solie; Bernice Sutliff; Clarence Werner.
Interviews by Don Franzman: Bob Donaldson; Irma Guettinger; Richard Kopp; Harvey Stabenow.
Interviews by Susan Glenz: Jeanne Kysilko Andre; Sally Hayden.
Interviews by Cathy Inderberg: James and Esther Bandli; Audrey Erickson; Harris Kahl; Carol and Larry Moen; Edwin and Merle Sjostrom.
Interviews by Harold Kringle: Mary and Vince Jesse; Carl Penskover; Vernon Peterson; Emil Sarauer.

Interview by Meg Marshall: Dan Riley
Interviews by Paul Nagel: Ed Bowe; Don and Margaret Gilbertson; Joe Laufenberg; Ben Rosenberg.
Interview by Jeanne Nyre and Carrie Ronnander: Harold Stahlbusch.
Interviews by Edward Pond: La Verne and Beverly Ausman; Tim Dotseth; Eugene Felix; Truman Felix; Alvin Kohlhepp.
Interviews by Mary Ellen Stolder: Tim Bandli; Dorothy Bullis Carpenter; Bruce Donaldson; Gary Evans; Nancy Frank; Judy Gilles; Carol and Harold Kringle; Don and Ilene Moos; Jan Morrow; Bill Nelson; Charlie Price; Steve Siverling; Kevin Splett; Leonard Splett; Nathan Splett; Dennis Stuttgen; Mike Taft; Dan and Jan Weiss.
Interviews by Frank Smoot: George Johnson; Larry Macke; Bev Peterson; Josephine Trunkel.
Interviews by Carol Wick: Esther Harriman; Menda Kopp; Ragnhild Preston; Delores Swoboda; Dan and Mary Emmerton.

Learning by Doing

Interviews by Bill Giese: Carol Bakke; Erma Quilling.
Interview by Steve Pederson: Marvin Barneson.
Interview by Julie Toske: David Schmid and Karen Bumann, Sweetland Farms.
Interviews by Jim Zak: Harriet Kovacezich; Mary Smith; Ruth Peterson; Jean Sieck.

Rural Life Documentation Initiative

Interviews by Janet Gilmore, Ph.D., folklorist: Esther and James Bandli; Darlene Honadel; Inez and Roger Robertson; Esther Schrock; Elaine Schroeder; Lucille Schumacher; Margaret Severson.
Interviews by James Leary, Ph.D., folklorist: Bud Epp; Mike Hable; Ralph Herman; Boniface "Billy" Jasicki, Rhythm Playboys; Joe Juza; Ralph Rubenzer, Rubenzer Hotshots; Chuck Schwab, Pines Ballroom; Ralph Sokup; George Theis, Our Saviour's Lutheran Church, Rice Lake.
Interviews by Ruth Olson, Ph.D., folklorist: Douglas Graff; Harold Graff; George and Sally Hayden; Douglas, Jane, Margaret, Peter and Steven Mueller; Jeff and Marie Pagenkopf; Walter Pagenkopf.

Archives and Collections: Research Assistance, Unpublished Information, Artifacts, Images and Documents

Heather Muir and Rita Sorkness, Area Research Center, University of Wisconsin-Eau Claire; Barron County Historical Society; Toni Kenealy, Cadott Historical Society; Cathy Mousette, Chippewa Valley Electric Co-op; Chippewa County Historical Society; Chippewa Falls Museum of Industry & Technology; Chippewa Valley Electric Co-Op; Dunn County Historical Society; General Mills Archives; Denise Zimmer, Jump River Electric Cooperative; Stacy Leitner, National Agricultural Statistics Service Information, U.S. Department of Agriculture, Wisconsin Office; Rusk County Historical Society; Bev Hompe, Stanley Area Historical Society; Bernard Schermetzler, University Archives, Steenbock Library, University of Wisconsin; Cathy Statz, Wisconsin Farmers Union; Wisconsin Historical Society.

Research and Collections Development Assistance

Marty Rugotzke, Augusta School District; WFLA/ZCBJ Bohemian Lodge Number 141, Cadott; Mike Hable, Bohemian Ovens, Bloomer; Mr. and Mrs. Harvey (Mary) Borntreger, Augusta; Staff, *Country Today*; Mary Kay Brevig, Eau Claire Energy Co-op; Staff, *Eau Claire Leader-Telegram*; Aaron Ellringer; Huntsinger Farms/Silver Spring Gardens, Eau Claire; Randy Knapp; Anthony Kotecki Family; Marv Lansing; Peggy Leum, Organic Valley/CROPP; Meg Marshall; Margaret Mueller; Scott Heiberger and Barbara Lee, National Children's Center for Rural and Agricultural Safety; Alaina Kolosh, National Safety Council; Beverly Peterson; Mahlon Peterson; Norma Pire; Mark A. Purschwitz; Tom Quinn; Rock Elm Methodist Church; Joan Roubal; Walt Schaffer; Therese Trojak; Arlene Verdegan; Perry Baird, Wisconsin Energy Cooperative News; Wisconsin Department of Transportation, District 6, Eau Claire.

Artifacts, Images and Documents:

Arvid and Linnea Anderson; Jeanne Andre; Joseph Antolak; Bob Artley; Rosey Asher; La Verne Ausman; Esther Bandli; Lois Battles; Donovan Bensend; Lenore Berg; Stacy Berger; Bethlehem Lutheran Church Ladies Aide, Ludington; Big Elk Creek Lutheran Church; Lois Bjork; Dennis Blodgett; Steve Boetcher; Carol and Cedric Boettcher; Mrs. Harvey Borntreger; Mary Bowe, Elizabeth and Wally Bowe; Mr. and Mrs. Bernard Brantner; Barb Brenden; Lila Brummond; John Butak; Dorothy Bullis Carpenter; Julie Carr; Chetek Lutheran Church; Gary Evans, Chippewa Valley Technical College; Colfax Lutheran Church; Mavis Cook; Bob Cropp; Sara Hawkins, D & D Dairy; DeWayne and Roxanne Dachel; Roland Dachel Family; Martha Davidge; Robert Donaldson; Herb Dow; Carl Erickson; Audrey Erickson; Steve Kinderman, *Eau Claire Leader Telegram/Country Today*; First Baptist Church, Eau Claire; Five Star Dairy; Carol Forster; Eric Fossum; Rolland Freid; Judy Gilles; Mary Gladitsch; Doug Graff; Harold Graff; Esther Harriman; Eugene Hayden; George and Sally Hayden; Ralph Hermann; Mr. and Mrs. Myron Hesselink; Lucille Hogstrom; Darlene Honadel; Mary Ann Huff; Ralph Ingman; Iowa Public Television; George Johnson; Jean Kannel; Troy Knutson; Elmer and Margery Kohlhepp; Ronald Kohnke; James Krings; Lois Krumenauer; Gerald and Sylvia Lee; Library of Congress, Rare Book and Special Collections Division; Library of Congress, Prints & Photographs Division; Ruth Lilly; Therese Olson, Lowes Creek Tree Farm; Lori Lyons; Ken and Kevin Mahalko; Janet and Wayne Brunner; Midvalleyvu Farm; Joan Mohr; Doug and Jane Mueller; Kathleen Mueller; Dave and Jeanne Nyre; Irene Ovren; Donald Parkhurst; Genevieve Pederson; Lori Peterlik; Beverly Peterson; Norma Pire; Charles "Chuck" Schwab, Pines Ballroom; Jennifer Schwab, Pines Ballroom; Barbara Bender, Prairie Farm Methodist Church; LeAnn Ralph; Sam and Alice Retallick; Inez and Roger Robertson; Dave and Joan Roubal; John Russell; Ralph Rubenzer; Marty Rugotzke; Reverend Hoeser, St. Henry's Catholic Church, Eau Galle; St. Joseph's Catholic Church, Elk Mound; St. Raymond of Peñafort Parish; SBC Communications; Esther Schrock; Lucille and Pete Schumacher; Jennifer Schwab; Margaret Segerstrom; Joe Simerdiek; Mary Jane Singerhouse; Kim and Steve Siverling; Dan Smetena; Smithsonian Institution; Kenneth and Virginia Smoot; Pam Solberg; Ralph and Augusta Sokup; Harold Stahlbusch; Diane Stasko; Sally Steinke; Arthur and Lenore Swan; Paul Swoboda; George Theis; Rose Tomesh; American Studies Group, The University of Virginia; Dave Voegeli; Allen and Phyllis Wathke; Katie Watters; Clarence Werner; Carol Wick; Winkler Prosthetic Labs, Eau Claire; Sylvan Winkler; Wisconsin Department of Transportation; Ken Ziehr.

Contributed Materials and Services

Barry and Mary Anderson; Ayres Associates; Kevin Dallas; Frances Jeche Interior Design; Renee Jacobson; Julie Johnson, artist; Lindstrom Equipment; Lyman Lumber; M & J Painting; Kris MacCallum, artist; Bill Musser; N.E.I.; David and Donna Nyre; Oak Ridge Chemicals; Mark Smith; Art Department, University of Wisconsin-Eau Claire; Geography Department, University of Wisconsin-Eau Claire; Walters Buildings, Augusta; David Wright; Ken Ziehr, architect.

Suppliers

Blue Rhino; County Materials; DigiCopy; Market & Johnson; Great Big Pictures; Lyman Lumber; Sharp One Hour Photo, Gail Schellinger, artist.

Learning by Doing

Learning by Doing provides graduate level professional development for classroom teachers. Fellows selected for the program learn public history methods and contribute to public history projects as part of their curriculum. Supported by major grants from the Department of Education, *Learning by Doing* is offered through a partnership between the Center for History Teaching and Learning at the University of Wisconsin-Eau Claire, the Chippewa Valley Museum, and Cooperative Educational Services Agency (CESA) 10. 2003-05 team projects contributed to development of *Farm Life*.

Fellows: William Giese, Shana Heidtke, Austin Herrick, James Jeffries, Christine Kadonsky, Steve Pederson, Brian Phelps, Julie Toske, James Zak.

Program Instruction and Coordination: University of Wisconsin-Eau Claire: John Mann, Jane Pederson, Roger Tlusty, Patricia Turner; Chippewa Valley Museum: Janet Dykema, Dondi Hayden, Carrie Ronnander, Susan McLeod, Jeanne Nyre, Frank Smoot, Eldbjorg Tobin; CESA 10: Joseph Mauer.

Interns

University of Wisconsin-Eau Claire: Aubrey Ashenbrener, History and Public Relations; Joshua Molnar, History; Stephen Sydow, History Graduate Program; Benton Ward, Art and Design.

University of Wisconsin-Milwaukee: Heather Ann Moody, Museum Studies Graduate Program.

Volunteers

Robert Accola; Arnie Anderson; La Verne Ausman; Roger Barstad; Amy Benson; Bill Benson; Candace Brown; Jon Case; John Chermack; Joe Ciurro; Kevin Dallas; Mark DeRusha; Jayne Dooley; Almira Downs; Katie Ediger; Bruce Flinn; Matt Gitzlaff; John Glenz; Susan Glenz; Len Grecco; Neil Green; Leah Hamann; Tom Harvey; Darcy Hayden; Gene Hayden; Arnold Heck; Ken Heilman; Brent Hensel; Lucille Hogstrom; Carolyn Huhmann; Ron Kahnke; Carolin Kauten; Eric Kayser; Jerry "Random" Kjesbo; Ron Kohnke; Max Lindoo; Steve Mann; Paul Marcello; Kathie Matter; Becky Mattson; Josh Mattson; E. J. "Buzz" McAdams; Dick Moore; Katie Nelson; Dave, Samantha, Carley and Kayla Nyre; Lars Olson; Signey Olson; Therese Olson; Kristin Pearson; Erin Pevan; Norma Pire; Chris Poppe; Gennie Prock; Sam Retallick; Marjorie Rice; Gail Schellinger; Tony Schuster; Kaye Senn; Roger Shepler; Sylvia Sipress; Pat Smith; Larry Sternitzky; Tom Stevens; Adam Stokstad; Jeff Thomas; Marcus Utzig; Donna Vierbicher; Jake Walter; Tom Walters; Andy Welsh; Danielle Wendt; Shari White; Carol Wick; Ken Ziehr.